T0295349

Smart Cities

Smart Cities: Blockchain-Based Systems, Networks, and Data examines the various components that make up a smart city. It focuses on infrastructure, processes, and services and outlines approaches for services such as health, transport, energy, and more. With an underlying emphasis on blockchain networks, the authors examine ways to provide the management of resources and activities by creating a more secure and trustless operating system where resources are more effectively allocated and managed.

Features include:

- Novel approaches toward the provision of smart city services
- Detailed explanations of how a blockchain-based smart city network operates
- Novel design and architecture for cutting-edge technologies such as energy systems and vehicular devices interacting with blockchain across smart cities
- Monitoring of data flow and the movement of several data types across different components of a smart city
- Comprehensive analysis of issues affecting entities across a smart city and the effects of blockchain-based solutions

This book is a practical and detailed demonstration for researchers and industry professionals who would use blockchain technology for effective city management.

Smart Cities
Blockchain-Based Systems, Networks, and Data

Jianbin Gao, Qi Xia, Kwame Omono Asamoah, and Bonsu Adjei-Arthur

CRC Press
Taylor & Francis Group
Boca Raton London New York

CRC Press is an imprint of the
Taylor & Francis Group, an **informa** business

First edition published 2023
by CRC Press
6000 Broken Sound Parkway NW, Suite 300, Boca Raton, FL 33487-2742

and by CRC Press
4 Park Square, Milton Park, Abingdon, Oxon, OX14 4RN

CRC Press is an imprint of Taylor & Francis Group, LLC

© 2023 Taylor & Francis Group, LLC

Library of Congress Cataloging-in-Publication Data

Names: Gao, Jianbin, author. | Xia, Qi, author. | Asamoah, Kwame Omono, author. | Adjei-Arthur, Bonsu, author.
Title: Smart cities : blockchain-based systems, networks, and data / Jianbin Gao, Qi Xia, Kwame Omono Asamoah, and Bonsu Adjei-Arthur.
Description: First edition. | Boca Raton : CRC Press, 2022. | Includes bibliographical references. | Summary: "Smart Cities: Blockchain-Based Systems, Networks, and Data examines the various components that make up a smart city. It focuses on infrastructure, processes, and services, and outlines approaches for services such as health, transport, energy, and more. With an underlying emphasis on blockchain networks, the authors examine ways to provide the management of resources and activities by creating a more secure and trustless operating system where resources are more effectively allocated and managed"-- Provided by publisher.
Identifiers: LCCN 2022014202 (print) | LCCN 2022014203 (ebook) | ISBN 9781032265575 (hardback) | ISBN 9781032266824 (paperback) | ISBN 9781003289418 (ebook)
Subjects: LCSH: Smart cities. | Blockchains (Databases) | Infrastructure (Economics) | City planning--Data processing.
Classification: LCC TD159.4 .G36 2022 (print) | LCC TD159.4 (ebook) | DDC 307.760285--dc23/eng/20220708
LC record available at https://lccn.loc.gov/2022014202
LC ebook record available at https://lccn.loc.gov/2022014203

ISBN: 978-1-032-26557-5 (hbk)
ISBN: 978-1-032-26682-4 (pbk)
ISBN: 978-1-003-28941-8 (ebk)

DOI: 10.1201/9781003289418

Typeset in Palatino
by KnowledgeWorks Global Ltd.

Contents

Preface

In recent times, there has been a migration from traditional means of settlement to modern forms of settlement and structures known as smart cities. This settlement incorporates major technological innovations as well as services to better enhance the lives of individuals and structures. The multifaced nature of smart city increases and manages successfully the complexity of organizing human settlement due to various components that must come together to ensure the sustenance of smart cities. Smart cities consist of several layers that are built on top of each other and must work together to ensure the achievement of various tasks and processes that take place in them. In normal traditional settlements, services and processes exist differently from infrastructure and individuals. All these are brought together to interact within a smart city.

Also, there exist several sectors such as governance, health, transportation, and business entities that control the daily processes within smart cities. These institutions as well as sectors run within their own sub-units and are thus not able to interact with each other without a trusted third party. Thus, several centralized units or agents are created, which help to ensure trust and enforce protocols set within the various units all across the smart city.

Again since these components remain within their own sub-unit, there exists a low degree of inter-operability for all these components within the smart city. This affects all processes that take place since an effective management system and control are needed to ensure that they function well. Since various components within the smart city depend and rely on resources shared by multiple units for decision-making and enactment of procedures, it is important to understand the smart city in full detail and provide the necessary solutions that ensure security, transparency, intractability, and autonomy for all components working within the blockchain. We believe the blockchain with its inherent features provides such means by which the various entities within the smart city can be able to operate successfully in a trusted manner. However, there is a need to understand the smart city in detail to provide a means by which the blockchain can best serve this need. We delve through this idea by proposing a blockchain-based smart city operating system that brings all the resources within a smart city together through the underlying blockchain network and provides a means by which all entities can exist, transact, and function in a more transparent and trustless manner.

Acknowledgment

This work was partially supported by the Sichuan Science and Technology Program (2019YFH0014, 2020YFH0030, and 2020YFSY0061).

Authors

Dr. Jianbin Gao earned his PhD in computer science from the University of Electronic Science and Technology of China (UESTC) in 2012. He was a visiting scholar at the University of Pennsylvania, Philadelphia, US from 2009–2011. He is currently an Associate Professor at the University of Electronic Science and Technology of China (UESTC).

Dr. Qi Xia is a Professor at the University of Electronic Science and Technology of China (UESTC). She is currently the deputy director of the Cyberspace Security Research Centre, the executive director of the Blockchain Research Institute, the executive director of the Big Data Sharing and Security Engineering Laboratory of Sichuan province, a member of the CCF blockchain committee, and a chief scientist at YoueData Company Limited.

Dr. Qi Xia serves as the PI of the National Key Research and Development Program of China in Cyber Security and has overseen the completion of more than 30 high-profile projects. She has published over 50 academic papers and won second place at the National Scientific and Technological Progress Award in 2012. Dr. Xia earned her BSc, MSc, and PhD degrees in computer science from the University of Electronic Science and Technology of China (UESTC) in 2002, 2006, and 2010, respectively. She was a visiting scholar at the University of Pennsylvania, Philadelphia, US from 2013–2014.

Her research interests include network security technology and its application, big data security, and blockchain technology and its applications.

Kwame Omono Asamoah earned a BSc in computer science from the Kwame Nkrumah University of Science and Technology, Ghana in 2014. He earned his master's in computer science and technology from the University of Electronic Science and Technology of China in 2018. Currently, he is pursuing a doctorate degree in computer science and technology at the University of Electronic Science and Technology of China. His current research includes blockchain technology and big data security.

Bonsu Adjei-Arthur earned his BSc in information technology from University of Ghana, Ghana in 2017, and an MEng in computer science and technology from UESTC, China in 2020. He is currently pursuing his PhD in computer science and technology in UESTC. His research interests include blockchain technology and security and privacy.

1

Introduction

1.1 Approach to Development of Smart Cities

In the most general sense, the term city refers to a large human settlement where there are structures from governance and population management to the provision of [1] services such as housing, transportation, sanitation, communications, and healthcare. These amenities make the city livable and assist residents as they go about their daily lives. Recently, however, the increase in rural-urban migration has seen a significant increase all over the world [2]. As a result, more people live in cities now than in any time in human history. Indeed more than 50% of the current global population lives in cities and urban centers. This number is expected to rise higher in the coming decades. This trend presents several benefits as well as challenges that must be managed carefully and speedily in order to maintain control over cities. On the bright side, more people in a given space means greater collaboration for business, more opportunities for job creation and an overall positive impact on the economy concerned. However, the dense concentration of people in these places puts severe pressure on available resources [3]. These pressures, when not properly managed, can have detrimental effects on the locale as competition for resources increases the prices of basic goods and services creating a social stratification that is unacceptable to some residents and fuels crime.

Several cities in Europe, North America, and Asia have already experienced the effects of the intense competition for city resources as more of their residents move from their rural dwellings into the cities [4, 5]. To better manage the service requirements of their populations, the authorities of these cities have increasingly turned to technology, applying research findings to modernize every aspect of the city. They have inspired and funded research into smart management of resources to leverage the reporting capabilities of sensors, and to mine the massive datasets generated for the maintenance of critical infrastructure. These sensors are networked to provide continuous data streams that indicate the structural health of various entities ranging from buildings, sewers, and storm drains to vehicles in transit, bridges, and cultural monuments. This increasing reliance on communications and networking technologies to monitor infrastructure and in some places provide

DOI: 10.1201/9781003289418-1

services with the data and analytics employed is what has given rise to the concept of the smart city. The definition of a smart city varies from one source to the other as the search for a commonly acceptable definition continues. This is due to the varying multiplicity of criteria each authority on the subject employed in their definition. However, for the purposes of completeness, the smart city is defined as a city that utilizes information and communications technology to assist the operations of the city and to provide services for its residents.

The smartest cities on the planet now are New York, London, Paris, Tokyo, Reykjavik, Singapore, Seoul, Toronto, Hong Kong, and Amsterdam [6, 7]. Others are incorporating more smart features into their management and operations in escalated bursts to grasp and better control the increasing complexity of city management. However, in many of the locales surveyed in the course of this work, smart cities typically identify one or two factors that mostly restrict the efficiency of their operations. They then proceed to install devices, the operations of which address the problem. For example, most cities have traffic jams as a constant factor that affects mobility. In many instances, they solve this either by widening the roads or by installing smart traffic management systems that can isolate and optimize the individual factors that reduce congestion on the roads [8]. The resulting improvement in the flow of traffic is then labeled a smart solution and hence the city is smart due to optimal flow of traffic. There are many challenges with this. First, it is immediately apparent that mobility is only one aspect of city dwelling and its efficiency has little or no effect on other equally critical factors such as healthcare, education, and sanitation. This challenge brings to the fore the common lack of integration in the research and technology of smart city development.

Cities are constantly being strained to provide energy, water, sanitation management, communications, healthcare, and many other services [7]. To meet the grave demand placed on city resources by its residents, it is essential to reconsider the services required to sustain life in the city, the infrastructure for the identified services and the efficient distribution of the resources. However, current trends attempt to service resource requirements in isolation leading to wastage and ineffective utilization with its attendant cascading negative effects [3]. To manage a smart city optimally through data-driven technologies necessitates the employment of a master resource manager. This resource manager must be dynamic to oversee the massive network of devices, sensors, data pools, and processes while maintaining control over multiple systems and heterogeneous networks, all of which must be achieved with high levels of abstraction. It must also integrate data into specific processes for targeted and specific interventions when and where required. To accomplish these goals, the smart city requires a smart city operating system, which is a tiered architecture whereby smart devices are grouped to form multiple subsystems till they constitute the emergent larger digital infrastructure capable of city management [9, 10].

1.2 Digital City Operating System – A Blockchain Perspective

It illustrates an adaptive and intelligent system of connected devices with an underlying mechanism for human behavior and motivation engineering to accomplish the goal of self-improving and reinforced reproduction of desired outcomes. Figure 1.1 depicts the architecture of the city operating system.

1.2.1 Users

As technology drives progress, human needs remain fundamentally the same: the necessities of food, shelter, health services, communication, and mobility have been with humankind since the beginning [9–11]. However, in the setting of a city where there are thousands of people and sometimes millions, these basic needs, which were perhaps easier to meet in small groups, assume challenging proportions. The competition for available resources and the inequalities in opportunities for knowledge, income, and other factors result in distribution problems, wastage, and other concomitant social problems. Attempts to solve these problems have resulted in the creation of economies supported by the provision of products and services such as food, transportation, and sanitation. [12]. Over the course of time, however, the need to create convenience in

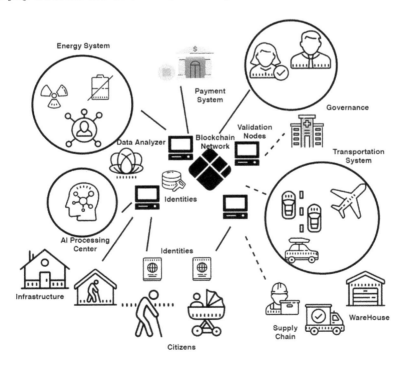

FIGURE 1.1
City operating system architecture.

oversight of management responsibility in the administration of the city leads to the centralization of resources and control by a select few fomenting several undesirable effects. The residents of the city who are required to keep the city running are left unable to partake fully in processes that benefit the city.

The smart city concept sets out to address these issues by refocusing attention on the reason for the city's existence, which is the residents themselves. By digitizing services rendered to residents and moving knowledge access to digital platforms, the smart city empowers its citizens with information and opportunities that can be leveraged to keep the city operating optimally. Through mobile access points (smart phones and tablets), residents may request and receive all or partial services. The smart city has the features listed next and these require one implementation or the other in its operating system:

1. *Smart-mobility-oriented transit*: Mobility in the smart city is informed by the availability of transport infrastructure, available capacity and intelligent decision-making on the best mode of travel in a given context.

2. *Smart infrastructure*: The supporting infrastructure for residents such as roads, houses, and parks supports active monitoring through extensive deployment of sensors, actuators, and transmission support equipment.

3. *Smart health*: Healthcare services in the city are proactive as opposed to the management of existing conditions. As such, there are widely deployed telehealth services for residents at each stage of patient health management.

4. *Smart tiered security*: Security is paramount in any transactional environment, and for a smart city, in order to maintain the security of people, the property should be protected in a manner that is effective while ensuring confidentiality, and avoiding unwarranted data access by external parties or influences.

1.2.2 Applications

The smart city invites residents to participate in its activities and processes by providing digital avenues for engagement. Every service that can be delivered by an online platform is made available on the Internet and most residents access these services using their smart phones [13]. The residents of Singapore check road traffic congestions by accessing the city's intelligent transport system (ITS) so they can save up to 60 hours annually. Digital health platforms are helping the citizens of London manage their health with mobile apps and telephone-based options that provide healthcare "anytime and anywhere." The city of Seattle saves approximately 700,000 gallons of water using advanced water supply monitoring sensor technologies. The instances cited previously highlight the extent of the adoption of smart technologies as well as actual tangible gains in economic and financial terms.

There are many such applications and they are usually focused around the following broad areas:

1. Identity
2. Transport
3. Sanitation
4. Healthcare
5. Supply chain
6. Security and privacy
7. Governance

These applications provide a core of services that underlie all transactions residents perform in the city.

1.2.3 Assistive Layer

1.2.3.1 Blockchain Analytics Engine

The smart city operating system is akin to a network operating system in that it is envisioned to be distributed and manages several thousands of devices of diverse heterogeneity networked to provide required services. The sheer volume of data to be collected from such devices can be reasonably expected to exceed available bandwidth [14]. As such, it is critical to develop a cooperative processing mechanism between sensor networks and the operating system to lighten the load of computation to actual necessity. An assistive layer will comprise a blockchain platform and a machine learning analytics engine that are recursively tuned to improve performance.

1.2.4 Hardware/Sensor Network

The Internet of things (IoT) is the empirical basis for constructing a smart city. IoT-based infrastructure for a smart city includes sensors that monitor buildings for parameters such as temperature fluctuations, humidity, and security of access to sensitive areas, water usage, and energy consumption among others [15]. Ultrasonic sensors installed in storm drains can determine the level of water in drains during and after storms and activate actuators to redirect the water elsewhere to avoid flooding or detect cracks in concrete infrastructure to alert crews in charge to carry out preventive maintenance. To achieve this in the most efficient manner, however, requires the sensors to be hierarchically networked and active. As such, the sensors monitoring conditions of the water pipes are grouped into sections for convenient management and all the sections together constitute a complete subsystem capable of independent action. Multiple subsystems then constitute another system depending on the resource under monitoring.

This design is intended to ensure only a minimum of reports from the sensors and actuators at the hardware layer reach the operating system. The benefits of such design are stated ad nauseam throughout this work. For practical purposes, the operating system is based on a network of networks that considers locales and contexts for data and processes. Typically, sensors in a building monitor parameters ranging from humidity to occupancy. In the design, all such sensors were grouped in a building under a single controller so that only reports of sensory activity are transmitted to the outside environment. Several buildings are grouped into a campus with a similar arrangement, electing a leader to oversee and coordinate activities. The campuses form a wide area network and these in turn constitute the distributed operating system for the smart city.

1.3 Conclusion

The smart city approach includes the use of smart solutions to improve the standard of living of residents, grow, and attract investment and promote environmental sustainability. The primary goal of a smart city is to maximize city functions and enhance economic growth while also improving people's standard of living through the use of smart technologies and data analysis. Technology's worth is decided by how it is applied rather than how much of it is available. This chapter described a strategy for creating a smart city. The digital city operating system was seen as the underpinning factor for proper smart city infrastructure management.

References

[1] D. Belanche, L. V. Casaló, and C. Orús, "City attachment and use of urban services: Benefits for smart cities," Cities, vol. 101, pp. 75–81, 2016.
[2] M. M. Rathore, A. Ahmad, A. Paul, and S. Rho, "Urban planning and building smart cities based on the internet of things using Big Data analytics," Comput. Networks, vol. 101, pp. 63–80, 2016.
[3] K. Su, J. Li, and H. Fu, "Smart city and the applications," in 2011 International Conference on Electronics, Communications and Control (ICECC), IEEE, 2011, September, pp. 1028–1031.
[4] M. Al-Hader, A. Rodzi, A. R. Sharif, and N. Ahmad, "Smart city components architecture," in CSSim 2009 – 1st International Conference on Computational Intelligence, Modelling, and Simulation, 2009.
[5] S. Maier, "Smart energy systems for smart city districts: case study Reininghaus District," Energy. Sustain. Soc., vol. 6, no. 1, pp. 1–20, 2016.

[6] M. Batty, K. W. Axhausen, F. Giannotti, A. Pozdnoukhov, A. Bazzani, M. Wachowicz, … and Y. Portugali, "Smart cities of the future," Eur. Phys. J. Spec. Top., vol. 214, no. 1, pp. 481–518, 2012.

[7] A. Caragliu, C. Del Bo, and P. Nijkamp, "Smart cities in Europe," in *Smart Cities*, Routledge, 2013, pp. 185–207.

[8] M. Castro, A. J. Jara, and A. F. G. Skarmeta, "Smart lighting solutions for smart cities," in Proceedings – 27th International Conference on Advanced Information Networking and Applications Workshops, WAINA 2013, 2013.

[9] K. O. Asamoah, H. Xia, S. Amofa, O. I. Amankona, K. Luo, Q. Xia, … and M. Guizani, "Zero-chain: A blockchain-based identity for digital city operating system," IEEE Internet of Things J., vol. 7, no. 10, pp. 10336–10346, 2020.

[10] R. M. Soe, "Smart twin cities via urban operating system," in ACM International Conference Proceeding Series, 2017.

[11] J. Frith, "Big data, technical communication, and the smart city," J. Bus. Tech. Commun., vol. 31, no. 2, pp. 168–187, 2017.

[12] G. C. Lazaroiu and M. Roscia, "Definition methodology for the smart cities model," Energy, vol. 47, no. 1, pp. 326–332, 2012.

[13] A. Zanella, N. Bui, A. Castellani, L. Vangelista, and M. Zorzi, "Internet of things for smart cities," IEEE Internet Things J., vol. 1, no. 1, pp. 22–32, 2014.

[14] K. Biswas and V. Muthukkumarasamy, "Securing smart cities using block-chain technology," in Proceedings – 18th IEEE International Conference on High Performance Computing and Communications, 14th IEEE International Conference on Smart City and 2nd IEEE International Conference on Data Science and Systems, HPCC/SmartCity/DSS 2016, 2017.

[15] F. Li, J. Hong, and A. A. Omala, "Efficient certificateless access control for industrial Internet of Things," Futur. Gener. Comput. Syst., vol. 76, pp. 285–292, 2017.

2

Blockchain and Smart City Fundamentals

2.1 Introduction

Blockchain is a decentralized, record of transactions that simplifies the recording of transactions and asset management in a business network. A single asset can be both tangible and intangible. A tangible asset (such as a house, car, cash, or land) and an intangible asset (such as a business) are two types of assets (examples include intellectual property, patents, copyrights, and branding) [1]. A blockchain network can track and trade almost anything of value. Information is the lifeblood of any business. The faster and more precisely it is received, the better. Blockchain is ideal for delivering that information because it provides instant, shareable, and completely transparent data stored on an immutable ledger that can only be read by network members with permission [2]. A blockchain network could be used to track orders, payments, records, manufacturing, and a variety of other things. You can see all facts of a transaction from beginning to end because members share a single view of the truth, giving you more sense of trust as well as fresh efficiencies and possibilities [3].

While blockchain technology has only recently been connected to new methods of dealing with capital assets, its capacity is virtually limitless. Nevertheless, in this revolutionary movement, broadening information transparency introduces new challenges and legalities, especially for security-conscious people [4]. In this chapter, we will discuss the basics of blockchain technology, and the useful aspects of what constitutes a blockchain and the innate dangers of a decentralized network in reality. Understanding blockchain technology is similar to working with a Google Doc. When we create a document and share it with a group of people, it is dispersed rather than replicated or moved. This results in a decentralized distribution chain in which everyone has simultaneous access to the document. No one is locked out while another party modifies the document, and all changes are tracked in real time, making them completely transparent. Blockchain is a particularly promising and revolutionary technology because it reduces risk, eliminates fraud, and enables scalable transparency

---

DOI: 10.1201/9781003289418-2

for a wide range of applications. Blocks, nodes, and miners are the three main concepts of blockchain.

2.1.1 Blocks

Every chain is composed of a number of blocks, each of which contains three basic components: the following are examples of the data found in the block. A nonce is defined as a 32-bit number [5]. A random nonce is generated when a block is built, and this nonce is then used to generate a block header hash. The nonce is used in conjunction with a hash, which is a 256-bit value. It must begin with a large number of zeros (i.e., be extremely small). A nonce generates the cryptographic hash when the first block of a chain is created. Unless mined, the data in the block is considered signed and irreversibly linked to the nonce and hash.

2.1.2 Miners

Mining refers to the process by which miners add new blocks to the chain [6]. Every block in a blockchain has its own unique nonce and hash, but it also refers to the previous block's hash, making block mining difficult, especially on large chains. Miners use specialized software to solve the mathematical problem of generating an acceptable hash using a nonce. Because the nonce is only 32 bits long and the hash is 256 bits long, there are approximately 4 billion nonce-hash combinations to search through before finding the correct one. When this occurs, miners are said to have found the "golden nonce," and their block is added to the chain. Any change to a block earlier in the chain necessitates re-mining not only the affected block, but also all subsequent blocks. This is why it is so hard to manipulate blockchain technology. Because finding the golden nonce takes a long time and a lot of computer resources, think of it as "safety in math." When a block is accurately mined, all network nodes affirm the change, and the miner is incentivized monetarily [7]. Furthermore, each block in this network usually has two hash values: one determining the block's identity and the other determining the block's unique parent (i.e., the previous block). There are no parents in the genesis block, which is the first block in the blockchain.

2.1.3 Nodes

One of the most important aspects in blockchain technology is decentralization. A single computer or organization cannot own the chain. Instead, the nodes that connect to the chain form a distributed ledger [8]. A node is any form of technological device that saves copies of the blockchain and keeps the network running. Every node has its own copy of the blockchain, and the network in order for the chain to be updated, trusted, and confirmed must approve any freshly mined block algorithmically. Because blockchains are transparent, every action on the ledger can be easily inspected and investigated. Each participant is given a unique alphanumeric identification number that is used to track his or her transactions [9, 10]. The blockchain's integrity is maintained and users' trust is built by combining public data with a system of checks and balances. In a nutshell, blockchains are the scalability of trust through technology.

2.1.3.1 Types of Blockchain Nodes

The line between node and non-node roles in a blockchain network is blurred. In some cases, multiple node types are possible. For example, Hyperledger allows for extensive role specialty, enabling nodes to focus on their strengths. Full and light nodes are two notable blockchain node differences [8]. Full nodes, as their name implies, do everything a node does. They maintain a complete ledger and help form blocks and reach agreement. A blockchain network's security and decentralization depend on a critical mass of full nodes. Lite nodes are aimed at facilitating transaction execution and validation without requiring full node services. In order to verify blockchain integrity, you only need the block headers [11]. Lite nodes only ask for transaction data when they want to verify a transaction's inclusion in a block. Decentralization is sacrificed for reduced storage and interaction regulations of lite nodes.

2.1.3.2 Blockchain Nodes Security

Nodes are the target of the majority of assaults against blockchain networks. While other attacks (such as 51% attacks) have a bigger name recognition, many attackers have learned that targeting individual users is more profitable [12]. Malware, phishing, and security flaws are all threats at the node level. Users make modifications to their blockchain software settings without fully understanding the repercussions, which leads to misconfiguration problems. For example, a famous Ethereum customer permits third-party applications to interact with wallet software via Remote Procedure Call (RPC). Attackers targeting port 8545 were able to connect and steal $20 million in Ether. Phishing scams aimed at blockchain users are also common. The Electrum wallet is infamous for being a phishing target, with a single attacker stealing over $1 million in Bitcoin in just a few hours.

Finally, blockchain node malware has many uses. A node's blockchain software can be targeted by malware for many of the attacks described in this series [13]. You have complete control over the security of a blockchain node if you run one. Taking the required security precautions, such as installing antivirus software, properly configuring it, and being careful of phishing scams, can make a huge difference in the security of both you and the blockchain network. While decentralization makes some vulnerabilities more difficult to defend against, each secure node actually adds to the network's total health and security.

2.2 Fundamentals of Blockchain

At its most basic level, blockchain technology is made up of cryptographic algorithms. Satoshi Nakamoto, the blockchain's creator, designed a system in which we place our faith in the blockchain and the cryptographic processes it employs rather than entities that keep trusted records (such as banks).

2.2.1 The Blockchain's Cryptography

Decentralized, distributed, and credible history records are the goal of the blockchain. Users can send and receive payments on the Bitcoin blockchain while remaining confident that their assets will not be lost or stolen. The blockchain relies on a few cryptographic techniques to achieve this level of trust. The blockchain ecosystem relies on hash functions and public key cryptography to function [14].

2.2.2 Hash Functions

A hash function is a mathematical function that accepts any integer as input and returns a number that falls within a specific range of values. Hash methods with a value of 256 bit, for example, give outputs in the range of 0–256 bits. To be deemed secure, a hash function must be collision-resistant; this means that finding two inputs that yield the same hash output is extremely difficult (almost impossible). The following features are required:

1. The hash function is free of defects.
2. There are numerous output possibilities.
3. A hash function that only goes one way (in which the input cannot be deduced from the output).
4. The outcomes of the same inputs are substantially different.

If a hash function meets these criteria, it can be utilized in blockchain. The blockchain's security is threatened if any of these rules are breached.

Blockchain primarily relies on cryptographic hash techniques to ensure that transactions can't be changed after they've been recorded in the ledger.

2.2.3 Public Key Cryptography

Public key cryptography [15] is another cryptographic technique utilized in blockchain technology. Because of its many advantageous qualities, this sort of encryption is commonly used on the Internet. Public key cryptography can be used to:

1. Encrypt a message so that it can only be read by the designated recipient.
2. Make a digital signature that proves you sent a certain communication.
3. Use a digital signature to confirm that a communication has not been tampered with while in transit.

Everyone has two encryption keys in public key cryptography: one secret and one public. Your secret key is a number that you choose at random. It's capable of deciphering messages and generating digital signatures. Your public key is created from your secret key and is intended for public distribution, as the name implies. It encrypts and generates digital signatures for messages that are sent to you. The majority of the time, your public key is utilized to generate your blockchain address (the address to which individuals send transactions).

Because of two factors, public key cryptography is secure. The first is the confidentiality of your private key [16]. Your blockchain account is completely at the power of anyone who can guess or steal your private key. This enables them to conduct transactions in your name and decipher data that is intended for you. People neglect to save their secret key, which is the most prevalent way for blockchain to be "hacked." Articles on quantum computers cracking blockchain are routinely reported due to the security of these "hard" problems [16]. Factoring and logarithms aren't difficult on quantum computers than multiplication and exponentiation, so classical public key cryptography is no more practicable. Nevertheless, since other challenges for quantum computers are still "hard," the quantum computer threat to blockchain can be mitigated with a simple upgrade.

2.2.4 Blockchain Construction

The blockchain is a collection of interconnected blocks that form a continuous whole, as its name suggests. We will look at how it works in this part. The blockchain's objective is to serve as a distributed ledger for safely storing data. These details are saved in the blocks. The block structure of a blockchain is depicted in the previous diagram. Throughout the series, we'll go over every aspect of this image, but for now, let's concentrate on the green parts. Each green component in the block represents a transaction. A transaction on a blockchain, such as Bitcoin, may reflect a literal transaction (i.e., a value transfer), but it is not the only option. Smart contract systems can also store information and send them as transactions, as we'll see in the following section. The security of public key cryptography determines the security of the blocks in the digital ledger.

2.2.5 Chaining

Every transaction and block is digitally signed by the blockchain's creator. Anybody with blockchain connection may simply verify that every transaction has been validated (i.e., submitted by somebody who possesses the associated address) and has not been modified after it was made. The digital signature of the block creator ensures the integrity and authenticity of the chain's blocks. Each block represents a single page in a bank's account ledger, representing only a portion of the network's history. Using hash techniques, the blockchain joins these slides into a continuous whole. The hash functions that connect each block are depicted in the previous diagram. Each block's block header (the area not carrying transaction data) carries the hash of the preceding block [17]. Because hash functions are collision-resistant, the fact that each block is dependent on the previous one is significant. Someone may either locate a different version of block 51 with the same hash or manufacture all of the blocks after 52 to generate the image's block 51.

2.2.6 Blockchain Networks

The blockchain is a decentralized ledger that can be trusted and shared. The network level is crucial when addressing the blockchain ecosystem since this ledger includes the blockchain network's history. In the previous post, we discussed the nodes and how they each keep their own copy of the distributed ledger. No other node will implicitly trust any other node's copy since the blockchain is supposed to be trustless. They will require a way to communicate over the network and a mechanism to agree on the ledger's present state (consensus).

2.2.7 Blockchain Peer-to-Peer Network

The network architecture of blockchains differs from that of typical online apps. The server operates as a single source of truth for these services, and clients connect directly to it to post or get application data [18]. Your email does not go immediately from your computer to the receiver when you use a webmail client like Gmail.

Instead, you send it to Gmail servers, where it is downloaded and read by the recipient after it has been downloaded. This method is simple and effective, but it relies on the Gmail server to operate as a trusted middleman. Blockchain relies on a peer-to-peer network in which each node connects directly with the others to avoid the need for trusted middlemen. In most blockchain networks, each message received by one of a node's five peers is relayed to the other four. Messages pass across the network in this manner via many paths, and no one has complete control over communications.

One key aspect of the peer-to-peer method to blockchain networking is that the underlying network must be capable of supporting it [18]. You can't adequately deploy a blockchain network over a network with varied levels of trust without jeopardizing blockchain or network security since each peer must be able to connect to every other peer. In addition, the blockchain's "broadcast" communication style requires a lot of bandwidth to work properly. Failure to support this could have a significant impact on the blockchain's security and effectiveness.

2.2.8 Blockchain Consensus

Blockchains are intended to be decentralized, dispersed systems. A big part of that is getting rid of the central authority that other systems depend on. Banks maintain control over the ledger, which shows how much money is stored in each account in a typical financial system. If there is a debate over which version of the ledger is the authoritative one, the bank has the final word. Blockchain's purpose is to decentralize centralized authorities such as banks. However, the system keeps a shared, decentralized ledger, with every node holding a copy and updating it as new blocks are formed. The difficulty here is guaranteeing that each block updates the ledger in the identical way across all nodes. Each block is created and shared by a temporary authority because the network lacks a permanent authority to establish the official version of the ledger. The means for doing so is the blockchain consensus algorithm [19].

2.2.9 Fundamentals of Consensus

The consensus algorithm's job is to keep the blockchain's control decentralized, which means that no single user can gain control of the network. This is accomplished by connecting the blockchain network's control to the possession of a valuable resource. Whatever consensus technique you use, the truth remains that control of a rare asset means power on the blockchain. This asset is computational power in proof of work. It is a blockchain-based proof of stake coin. The rationale for employing a scarce resource as a blockchain equivalent to power is that it allows incentive structures to be used to safeguard the blockchain. According to the Law of Supply and Demand, if there is a growing market for a resource with a limited supply, the price will rise.

An attacker must first get more of the valuable resource before attempting to seize control of a blockchain network (for example, by launching a 51% attack). As a result, they increase demand for the resource, driving up acquisition costs. The attacker's cost of obtaining enough of the resource to carry out a successful attack should, hopefully, be insurmountable. If not, successful 51% assaults on blockchains have been observed, most notably on smaller cryptocurrency networks.

2.2.10 Consensus Implementation

It was the only blockchain in existence when Satoshi Nakamoto built Bitcoin. The proof of work consensus algorithm employed on the Bitcoin network is explained in the whitepaper. Since then, a host of different consensus algorithms for a range of blockchain systems have been developed. Proof of Stake has received a lot of attention as a result of its inclusion in the Ethereum roadmap.

2.2.11 Proof of Work

Proof of work is the basic consensus algorithm, and it needs players to do work, as the term indicates. Miners are indeed the ones who try to make a new block in proof of work. The winner of a race in which the winner is the one who creates the block is designated as the block creator. This race comprises building a valid block with the condition that the block's header hashes to a value less than a certain threshold. Because of the nature of hash algorithms, random guessing is the most straightforward method. As a result, the network's miners experiment with different hashes until they find one that delivers the necessary hash output. The first miner to discover a valid block delivers it to the rest of the network, which utilizes it to generate the following block.

The biggest problem with proof of work is that building a valid block is the primary condition for creating one. Nothing prevents two miners from finding several copies of the block at the same time. If this happens, distinct parts of the network may grow on top of different blocks, resulting in a blockchain that diverges. Blockchain resolves issue using the longest block rule, which specifies that if two versions of the blockchain conflict, the longer one must be accepted. Proof of work, which employs the concept of difficulty, also aims to limit the chance of different blockchains. A valid block header's hash value must be less than a threshold value that can be adjusted in a distributed fashion. To ensure that blocks are made at the desired block rate, the difficulty is modified at regular intervals (based on the current processing power of the blockchain network) [20].

2.2.12 Proof of Stake

Proof of stake is a revolutionary method to blockchain security that uses a finite resource. Proof of stake uses the blockchain's unusual coin instead of restricted processing power (as in proof of work). Proof of stake is similar to investing in a business. You have the right to receive investor dividends if you provide a portion of your money to a corporation. In proof of stake, you promise not to spend (or stake) a portion of your Bitcoin in exchange for the chance to build a block (and earn the associated rewards). The techniques for selecting block creators based on stakes vary depending on the implementation. In fact, if a block maker is given two versions of the blockchain to build on, it is in their best interest to develop on both such that whichever version eventually takes over includes the block that pays them a block reward [21].

2.2.13 Consensus Attacks

The blockchain is controlled by consensus methods. As a result, a number of attacks have been launched in an attempt to seize control of the blockchain. An attacker can undertake a double-spend attack if they are successful, which allows them to finish one transaction and then remove it from the ledger later. Some consensus assaults have been around since the beginning (such as the 51% attack), while others (such as long-range attacks) have only recently been invented. Overall, 51% assaults, which occur when the blockchain's economic incentives fail, are the easiest way to attack a proof of work blockchain. When two versions of the blockchain are given, the longest block rule mandates every benign node to choose the longer one. An attacker who has complete control over the blockchain can build the lengthier version whenever they want.

Proof of work does this by controlling over half of the blockchain network's computing power. Because legitimate block generation requires a random search of plausible alternatives, the fastest searcher will be able to generate blocks. Similar assaults on proof of stake are conceivable, but require more control over the valued resource. In proof of work, you require 50% of the computing power to discover the next block. To forge the next block in proof of stake, you must control 100% of the staked coin. To challenge the consensus mechanism on proof of stake blockchains, long-range attacks can give an attacker control over a percentage of the staked cryptocurrency.

In this attack, the adversary makes a new blockchain that starts at the genesis block (this assumes that they have a stake in the genesis block). Choosing the attacker as the block creator produces a new block. They are the only ones who make blocks and hence collect block rewards. The adversary gains control of errant blockchain. To be recognized, the diverging blockchain must be longer than the "real" blockchain. A diverging blockchain will lag behind the main chain if a benign user is chosen to generate a block. A diverging blockchain will lag behind the main chain if a benign user is chosen to generate a block. The attacker's chain starts far behind, so this happens less frequently as they accumulate more stake.

2.3 Permissionless vs. Permissioned Blockchains

All types of blockchains can be described as permissionless, permissioned, or both [22]. In other words, permissionless blockchains allow anyone to join the network pseudo-anonymously (as "nodes") and have unlimited power. Permissioned blockchains, on the other hand, limit network access and network node power. A permissioned blockchain's users know the identities of other permissioned blockchain users.

Because there are more nodes to validate transactions, permissionless blockchains are more secure. Bad actors couldn't work together on a network. With so many nodes and transactions, permissionless blockchains have long transaction processing times. Permissioned blockchains are more efficient.

Having less nodes means faster transaction processing. In general, permissionless blockchains are more secure than permissioned blockchains. Bad actors couldn't work together on the network. Having a large number of nodes and transactions causes long transaction processing times. Permissioned blockchains are more efficient. Because network access is limited, the blockchain has fewer nodes and processes transactions faster.

2.3.1 Types of Blockchains

There are four different types of blockchain. These are as follows:

1. *Public blockchains*: Because public blockchains are permissionless and decentralized, anybody can participate. Public blockchains allow all nodes to connect to the network, make new blocks of data, and validate existing blocks of data. Most Bitcoin transaction and mining occurs on public blockchains. You may be familiar with Bitcoin, Ethereum, and Litecoin. Nodes construct blocks for the network's transactions by solving cryptographic equations. The miner nodes are rewarded with a little amount of Bitcoin. The miners, like modern bank tellers, create transactions and are rewarded (or "mined") for their work.

2. *Private blockchains*: Private or managed blockchains are permissioned blockchains regulated by a single entity. Nodes in a private

blockchain are chosen by a central authority. In addition, the central authority does not always provide each node the same permissions. Because they are private, they are only partially decentralized. These include Ripple, a B2B trading network, and Hyperledger, an open-source blockchain application framework. In addition to the time it takes to certify fresh data, private blockchains are more prone to fraud and unethical actors. Consortia and hybrid blockchains were created to overcome these challenges.

3. *Consortium blockchains*: Contrary to private blockchains, consortium blockchains are permissioned blockchains maintained by multiple organizations. In this way, consortium blockchains are more secure than private blockchains. Creating consortiums, on the other hand, can be difficult due to logistical issues and the risk of antitrust violations (which we will examine in an upcoming article). Blockchain consortia solutions from R3 are widely used in the financial services industry and beyond. A non-profit blockchain cooperative, the Global Shipping Business Network Collaboration aims to digitalize the shipping industry and improve collaboration among maritime companies.

4. *Hybrid blockchain*: Hybrid blockchains are administered by a singular body but are supervised by the public blockchain. IBM Food Trust is a hybrid blockchain designed to improve food supply chain efficiency. We'll get into IBM Food Trust in a subsequent piece in this series.

2.4 Smart Contracts

Blockchain technology was created to replace the banking system. The distributed, decentralized ledger of the blockchain network is used to record transactions. As a result, cryptocurrencies like Bitcoin can be used to create whole capital markets without a central bank (like a bank). Blockchain's global, decentralized ledger is beneficial for more than merely documenting financial transactions. On top of the blockchain, smart contract platforms operate a Turing-complete computer, allowing smart contracts to perform various tasks.

2.4.1 Introduction to Smart Contracts

Smart contract platforms use blockchain technology, but it is modified to allow third-party programs to run on top of it. Instead of actual money, transactions are computer instructions that the blockchain's virtual machine can execute. Simple "if/when … then …" lines in blockchain programming make smart contracts work. Once the prerequisites are completed and accepted, a computer network takes over. These tasks include transferring payments, registering a car, providing alerts, and issuing tickets. After the transaction, the blockchain is updated. The transaction cannot be modified, and the results are only available to those who have been granted access. As many parameters as needed to persuade participants that the activity will be completed correctly.

For this, participants must agree on how transactions and data are represented on blockchain, on the "if/when … then …" rules that govern those transactions, analyze any possible exceptions, and design an arbitration structure. There is no central computer system that runs the code and keeps track of the smart contract platform because the blockchain network is decentralized. Instead, each network node has its own virtual computer that runs the code in each blockchain transaction block. It is possible because code is predictable and divided into blocks before execution.

2.4.2 Smart Contracts Working Principle

In other words, smart contracts allow you to transfer assets or currencies. The software runs the code under the stated conditions. It does so automatically and utilizes the data to ensure a transaction's legality. It is activated once the condition is met. If the condition is not met, the smart contract is executed [23]. A decentralized ledger duplicates the smart contract or document for immutability and security.

2.4.3 Examples of Smart Contracts

Blockchain is used in the real estate sector. You can see smart contracts in operation if you buy real estate utilizing a blockchain-based platform in the future. Your first step is to acquire a house. Buying real estate involves many factors to consider. You will need to set loan amounts, payment dates, and other details. To activate a smart contract, you must first sign and set it in motion. Assume you agreed to pay 20% of the property's value up front. Then you agreed to pay the remaining real estate value in installments, plus any other conditions. The seller develops a smart contract based on the terms agreed. That is when the smart contract is activated. The smart contract will keep track of the payments. Once the vendor has received all funds, you possess the property. No middlemen or third parties are involved in any of the processes. Compared to a regular real estate transaction, you will save time and effort. Because there are no middlemen, both parties save money. Furthermore, in the event of a smart contract occurrence, all parties concerned will be notified.

2.4.4 Uses of Smart Contracts

Healthcare, supply chain management, and financial sectors all use smart contracts. Here are several examples:

1. *Government voting system*: Smart contracts make voting more safe by reducing tampering. Smart contracts' votes would be ledger-protected, making them harder to decode. Voter attendance has always been low due to an inefficient system that requires voters to queue, show ID, and fill out paperwork. Voting with smart contracts increases the number of participants in a voting system.

2. *Healthcare*: Blockchain can store patients' encrypted health records. Due of privacy considerations, only selected individuals would have access. Smart contracts can also be utilized to conduct secure and private research. The blockchain can keep all patient hospital receipts and automatically exchange them with health insurers. The ledger can be used for supply management, drug monitoring, and regulatory compliance.

3. *Supply chain*: Paper-based processes with several approval channels have traditionally hampered supply chains. The lengthy process increases fraud and loss risk. Blockchain can solve these problems by providing all participants with a secure and accessible digital version. Smart contracts can be utilized for inventories, payments, and task automation.

4. *Financial services*: There are many ways smart contracts help change traditional financial services. Then they compensate the user if everything is in order. Smart contracts eliminate the risk of tampering with accounting records. Shareholders can make transparent decisions. They also aid in trade clearing, which is the act of transferring funds after transaction settlement.

2.4.5 Advantages of Smart Contracts

1. *Secure*: Smart contracts guarantee safe contract execution. This protects the terms and conditions and other sensitive data. To implement smart contracts, no third-party or human involvement is required. To protect sensitive data, cryptography is used to encrypt it.

2. *Autonomous*: Smart contracts are self-contained, increasing their utility to new heights. After being begun, smart contracts can execute and complete themselves without human intervention.

3. *Interruption-free*: A third party cannot interrupt a smart contract that was never intended to be interrupted.

4. *Trustless*: Smart contracts secure all parties' interests in a trustless environment.

5. *Cost effective*: Smart contracts are cost-effective since they are self-contained and do not require an intermediary.

6. *Fast performance*: Smart contracts move quickly. Instead of hours, a contract can be finalized in minutes.

2.4.6 Limitations of Smart Contracts

1. *Difficult to change*: A smart contract's programming fault can take a long time and money to fix.

2. *Possibility of loopholes*: Parties will act in good faith and not profit unethically from a contract. Smart contracts, on the other hand, make it impossible to enforce the terms as intended.

3. *Third party*: However, smart contracts cannot completely eliminate third-party involvement. Third parties have a different function in traditional contracts. Lawyers, for example, will not be required to write individual contracts but will be required to help developers comprehend the requirements.

4. *Vague terms*: Smart contracts cannot always handle ambiguous terms and conditions since contracts demand unfamiliar vocabulary.

2.5 Blockchain Applications

Initially, the blockchain was designed to support Bitcoin (CRYPTO: BTC). Satoshi Nakamoto created an immutable record of transactions that links blocks of data using digital cryptography to solve the double-spending problem. However, blockchain technology may be utilized for much more than only Bitcoin and other cryptocurrencies. Here are a few.

1. *Money transfers*: The initial idea behind blockchain technology is still a great application. Traditional money transfer techniques may be more costly and time-consuming than blockchain. This is especially true for slow and costly international transactions. Money transfers between accounts can take days in the existing US financial system, but only seconds on blockchain.

2. *Real estate*: Property sales require a lot of documents to confirm custodianship and handover deeds and titles. Using blockchain technology to record property sales can increase security and accessibility. This reduces documentation and speeds up transactions, saving you time and money.

3. *Secure share of medical data*: Doctors and medical professionals can get accurate patient information by storing medical data on a blockchain.

This can help patients who see multiple providers get the best care possible. Some medical documents can be recovered faster, allowing for faster treatment. If insurance information is stored in the database, clinicians may immediately verify patient coverage and therapy reimbursement.

4. *Secure Internet of things*: However, the Internet of things (IoT) enables unscrupulous actors access to our data and control over critical infrastructure. Passwords and other data are stored on a decentralized network rather than a single server using blockchain technology. A blockchain also safeguards data from tampering due to its immutability.

5. *Supply chain and logistics tracking*: Using blockchain to track items as they move through a supply chain offers many benefits. First, data is stored on a secure public ledger, making collaboration easier. Second, the intransigence of blockchain data ensures data security and integrity. As a result, logistics and supply chain partners may work more freely knowing their data is true.

6. *Voting*: Blockchain technology may soon allow us to vote using personal identification data stored on a blockchain. Blockchain-based ballots cannot be manipulated, and no one can vote twice. Enabling voting using a smartphone can enhance voter turnout. Organizing an election would also be much cheaper.

7. *Government benefits*: A blockchain-based digital identity can also be used to manage government benefits like welfare, social security, and Medicare. Blockchain technology may cut fraud and operating costs. Meanwhile, blockchain-based digital disbursement speeds up payments.

8. *Artist royalties*: Artists can be paid for their work by using blockchain technology to track music and film files released online. Because blockchain technology was intended to prevent duplication of content, it can be used to combat piracy. Making royalties payments transparent and using a blockchain to track streaming platform playbacks will help artists be paid.

9. *Energy*: Blockchain technology may revolutionize the energy industry. Rooftop solar, electric vehicles, and smart metering have all boosted the energy sector. Due to its smart contracts and system compatibility, the Enterprise Ethereum blockchain is now marketed as the next big thing in the energy business. Using distributed ledger technology to track grid commodities can improve utility company operations. Aside from provenance monitoring, blockchain can help distribute sustainable energy.

10. *Satellite communication*: While blockchain is most commonly associated with the finance industry, it offers a wide range of potential

uses in space. Space industry participation in the development and broad acceptance of blockchain can be greatly aided by the deployment of satellites as nodes in the chain. Satellites can receive, store, and broadcast blockchain data and apps. To store data and process transactions, satellite networks serve as infrastructure.

2.6 Smart City and Blockchain Technology

Big data and the IoT have facilitated the establishment and viability of several smart city initiatives. It is a utopian urban development that uses information and communication technology to allow residents, governments, for-profit and non-profit organizations collect and exchange real-time data [24]. An urban innovation ecosystem aiming to overcome the issues associated with high urban population expansion by combining information and communication technology and applying next-generation ideas to all aspects of life. Human and social capital, traditional infrastructure (including transportation), and modern ICT infrastructure must be invested heavily in order to create sustainable economic growth, improve citizens' quality of life, and protect natural resources. An inclusive governance structure that emphasizes business-led urban development, high-tech and creative industries, social and relational capital, and environmental sustainability defines a smart city, according to Halevi et al. [16].

To overcome the challenges of urbanization, more and more communities are considering "smart cities" to improve residents' quality of life and support sustainability [24]. A multi-stakeholder, municipally based cooperation is required for smart cities, according to Manville et al. Also, the idea of smart cities is built on a shifting worldview. The seamless integration of ubiquitous digital technology with urbanization can improve many municipal services and residents' socioeconomic status. As a result, smart cities are enabled by technology. The goal of reducing the effects of rapid urbanization cannot be achieved without effective, real-time, and reliable data access and processing [25]. To manage resources efficiently, add high-value services, and improve residents' quality of life are all goals of smart cities. Interconnectedness among stakeholders allows for a rapid transition to smart cities, helped by technology. A variety of factors influence one's level of intelligence. Smart cities' structural components, such as infrastructure and services, must be compatible with the urban environment's design. According to Chourabi et al. [26], the essence of smart cities is identifying citizen needs and finding ways to address them.

Smart city designers should consider how current technology architecture might address economic, social, and environmental concerns. For enterprises to gain reliable business information, Lv et al. [27] claim that the growth of smart cities brings unprecedented challenges. Transparency and ease of use are crucial for smart city infrastructure such as transportation, electricity grids, healthcare, telecommunications, education, and government services. As an example, previous research has emphasized the need for smart city designers to create more secure networked relationships, interoperable data exchange models and efficient sharing economy platforms. In order to maintain a better infrastructure spanning many technical components and services, core technologies must be developed. Smart city projects must also increase service quality and handle various urban activities efficiently for smart residents. A potential driver of innovation, blockchain technology is an obvious choice for enabling a range of use cases in this regard. It may assist smart cities overcome difficulties related to the technological, social, and financial environments, as well as the absence of interaction between these factors.

2.7 Application of Blockchain in Smart Cities

We discovered many clusters of themes where the usage of blockchain technology might lead to advancement and make the building of smart cities easier. The sections that follow will focus on six of the most common blockchain application fields in smart cities. One example is smart healthcare. Other examples are smart supply chains and logistics; mobility; energy; e-voting; manufacturing; and education.

1. *Smart healthcare*: Blockchain ensures high-quality e-health services, which may assist deliver more personalized and effective treatment. Instead than being controlled by a single entity, blockchain provides for secure decentralization of patient data, making medical records more accessible to patients and professionals in an emergency. Thanks to instant access to health data, doctors may be able to identify severe conditions and save countless lives.

2. *Smart logistics and supply chain*: Blockchain technology has the potential to improve logistics and supply chain performance in smart cities. It can facilitate communication and information exchange among many stakeholders involved in logistics procedures. Supply chain stakeholders may use blockchain to more efficiently transport services and goods throughout the network, secure IoT devices, and streamline due diligence. Blockchain enables real-time tracking of all company transactions, providing value to the end products. Uncertainty and operational inefficiencies are reduced through a visible supply chain.

3. *Smart mobility*: Blockchain technology can help the automotive sector move to MaaS solutions such as remote software-based

vehicle maintenance, insurance, smart charging, and car-sharing. Decentralized ledgers enable a more efficient, safe, and flexible transportation system that balances the demands of present and future generations. Similarly, blockchain applications can support smart mobility by improving routing and scheduling, meeting citizens' requirements, and promoting environmental friendliness.

4. *Smart energy*: Blockchain's capacity to improve energy efficiency and resource planning and management can be immensely advantageous to smart cities. A recent study by Aggarwal et al. [28] found that blockchain can effectively regulate energy transformation and distribution in the smart grid, increasing transparency. By directly connecting energy resources and appliances, Park et al. [29] claim that customers will have access to high-quality, low-cost energy everywhere and at any time. Blockchain can help the energy network speed up and secure P2P energy transactions.

5. *Smart factories*: Blockchain can help smart factories be more reactive, adaptable, and ready to deal with unpredictable operational conditions. The technique is used for remote software upgrades, automatic fault diagnostics, and predictive maintenance. Blockchain can be used to operate and maintain industrial facilities that rely heavily on complex electro-mechanical equipment and can save significant amounts of money through predictive maintenance and material traceability. Ledger technology enables autonomous workflows and service sharing among industrial IoT devices.

6. *Smart home*: Incorporating blockchain technology into smart homes has many benefits. Blockchain can help with trust and traceability in smart homes, say Pacheco et al. [30]. On-demand service delivery is enabled by tracking the equipment and sensors utilized to offer it. Smartphone apps and IoT devices can readily collect data on user behavior, energy consumption, security, and human physiological data. Everything can be kept on the blockchain and shared economy services. Digital signatures can be used to identify suspicious activities and secure each smart home device's identity. The decentralized, transparent, and secure nature of blockchain makes it ideal for smart home sensors, actuators, and devices to communicate and share data.

2.8 Conclusion

Researchers have recently become interested in blockchain technology due to its intriguing properties. The blockchain technology can be used in smart city applications. Therefore, this chapter introduced blockchain and smart city. The smart city delivers smart systems that raise citizens' living standards. This chapter looked at how blockchain technology can be used in smart cities. The concept of a smart city was also examined in length. The chapter also discussed the merits and drawbacks of blockchain technology.

References

[1] Allee, V., 2008. Value network analysis and value conversion of tangible and intangible assets. *Journal of Intellectual Capital*, (9), pp. 5–24.

[2] Nofer, M., Gomber, P., Hinz, O. and Schiereck, D., 2017. Blockchain. *Business & Information Systems Engineering, 59*(3), pp. 183–187.

[3] Gamage, H.T.M., Weerasinghe, H.D. and Dias, N.G.J., 2020. A survey on block-chain technology concepts, applications, and issues. *SN Computer Science, 1*(2), pp. 1–15.

[4] Hylock, R.H. and Zeng, X., 2019. A blockchain framework for patient-centered health records and exchange (HealthChain): evaluation and proof-of-concept study. *Journal of Medical Internet Research, 21*(8), p. e13592.

[5] Shackelford, S.J. and Myers, S., 2017. Block-by-block: leveraging the power of blockchain technology to build trust and promote cyber peace. *Yale Journal of Law & Technology, 19*, p. 334.

[6] Calvão, F., 2019. Crypto-miners: digital labor and the power of blockchain technology. *Economic Anthropology, 6*(1), pp. 123–134.

[7] Chung, K., Yoo, H., Choe, D. and Jung, H., 2019. Blockchain network based topic mining process for cognitive manufacturing. *Wireless Personal Communications, 105*(2), pp. 583–597.

[8] Lin, I.C. and Liao, T.C., 2017. A survey of blockchain security issues and challenges. *International Journal of Network Security, 19*(5), pp. 653–659.

[9] Elrom, E., 2019. Blockchain nodes. In *The Blockchain Developer* (pp. 31–72). Apress, Berkeley, CA.

[10] Aoki, Y., Otsuki, K., Kaneko, T., Banno, R. and Shudo, K., 2019, April. Simblock: a blockchain network simulator. In *IEEE INFOCOM 2019 – IEEE Conference on Computer Communications Workshops (INFOCOM WKSHPS)* (pp. 325–329). IEEE.

[11] Chen, H.C., Irawan, B. and Shae, Z.Y., 2018, July. A cooperative evaluation approach based on blockchain technology for IoT application. In *International Conference on Innovative Mobile and Internet Services in Ubiquitous Computing* (pp. 913–921). Springer, Cham.

[12] Saini, H., Bhushan, B., Arora, A. and Kaur, A., 2019, July. Security vulnerabilities in information communication technology: blockchain to the rescue (a survey on blockchain technology). In *2019 2nd International Conference on Intelligent Computing, Instrumentation and Control Technologies (ICICICT)* (Vol. 1, pp. 1680–1684). IEEE.

[13] Liu, L. and Xu, B., 2018, April. Research on information security technology based on blockchain. In *2018 IEEE 3rd International Conference On Cloud Computing and Big Data Analysis (ICCCBDA)* (pp. 380–384). IEEE.

[14] Zeng, Z., Li, Y., Cao, Y., Zhao, Y., Zhong, J., Sidorov, D. and Zeng, X., 2020. Blockchain technology for information security of the energy internet: fundamentals, features, strategy and application. *Energies, 13*(4), p. 881.

[15] Hellman, M., 1978. An overview of public key cryptography. *IEEE Communications Society Magazine, 16*(6), pp. 24–32.

[16] Halevi, S. and Krawczyk, H., 1999. Public-key cryptography and password protocols. *ACM Transactions on Information and System Security (TISSEC)*, 2(3), pp. 230–268.

[17] Wang, J., Wu, P., Wang, X. and Shou, W., 2017. The outlook of blockchain technology for construction engineering management. *Frontiers of Engineering Management, 4*(1), pp. 67–75.

[18] Li, Z., Barenji, A.V. and Huang, G.Q., 2018. Toward a blockchain cloud manufacturing system as a peer to peer distributed network platform. *Robotics and Computer-Integrated Manufacturing, 54*, pp. 133–144.

[19] Bach, L.M., Mihaljevic, B. and Zagar, M., 2018, May. Comparative analysis of blockchain consensus algorithms. In *2018 41st International Convention on Information and Communication Technology, Electronics and Microelectronics (MIPRO)* (pp. 1545–1550). IEEE.

[20] Gervais, A., Karame, G.O., Wüst, K., Glykantzis, V., Ritzdorf, H. and Capkun, S., 2016, October. On the security and performance of proof of work blockchains. In *Proceedings of the 2016 ACM SIGSAC Conference on Computer and Communications security* (pp. 3–16).

[21] Kiayias, A., Russell, A., David, B. and Oliynykov, R., 2017, August. Ouroboros: a provably secure proof-of-stake blockchain protocol. In *Annual International Cryptology Conference* (pp. 357–388). Springer, Cham.

[22] Helliar, C.V., Crawford, L., Rocca, L., Teodori, C. and Veneziani, M., 2020. Permissionless and permissioned blockchain diffusion. *International Journal of Information Management, 54*, p. 102136.

[23] Omar, I.A., Hasan, H.R., Jayaraman, R., Salah, K. and Omar, M., 2021. Implementing decentralized auctions using blockchain smart contracts. *Technological Forecasting and Social Change, 168*, p. 120786.

[24] Biswas, K. and Muthukkumarasamy, V., 2016, December. Securing smart cities using blockchain technology. In *2016 IEEE 18th International Conference on High Performance Computing and Communications; IEEE 14th International Conference on Smart City; IEEE 2nd International Conference on Data Science and Systems (HPCC/SmartCity/DSS)* (pp. 1392–1393). IEEE.

[25] Nam, K., Dutt, C.S., Chathoth, P. and Khan, M.S., 2021. Blockchain technology for smart city and smart tourism: latest trends and challenges. *Asia Pacific Journal of Tourism Research, 26*(4), pp. 454–468.

[26] Chourabi, H., Nam, T., Walker, S., Gil-Garcia, J.R., Mellouli, S., Nahon, K., Pardo, T.A. and Scholl, H.J., 2012, January. Understanding smart cities: an integrative framework. In *2012 45th Hawaii International Conference on System Sciences* (pp. 2289–2297). IEEE.

[27] Lv, Z., Li, X., Wang, W., Zhang, B., Hu, J. and Feng, S., 2018. Government affairs service platform for smart city. *Future Generation Computer Systems, 81*, pp. 443–451.

[28] Kaushik, S., Aggarwal, G. and Tejasvee, S., 2021. A glimpse on key applications of smart city under M2M communication. In *IOP Conference Series: Materials Science and Engineering* (Vol. 1022, No. 1, p. 012006). IOP Publishing.

[29] Jeong, Y.S. and Park, J.H., 2019. IoT and smart city technology: challenges, opportunities, and solutions. *Journal of Information Processing Systems, 15*(2), pp. 233–238.

[30] Pacheco Rocha, N., Dias, A., Santinha, G., Rodrigues, M., Queirós, A., and Rodrigues, C., 2019. Smart cities and healthcare: A systematic review. *Technologies, 7*(3), p. 58.

3

Infrastructure

3.1 Introduction

A smart city is made up of people dwelling in various parts of cities, and as such, a smart city needs to contain a degree of state-of-the art facilities. These facilities should be able to connect both the physical, economic, and social parts of the city. The infrastructure necessary depends on the people living in the city at a particular time. There is a need to address infrastructure issues for all citizens, government, and planners to create a well-endowed smart city [1].

Understanding a smart city's infrastructure reflects the socioeconomic context. A contemporary city is a control system for assets and people. The infrastructure manages the flow of entities in the city such as service, asset, and even information. These infrastructures are fueled by Internet of things (IoT), Cloud, big data, and 5G mobile [2].

3.2 Smart City Infrastructure

Water, electricity, sanitation, solid waste management, urban mobility, and public transport are linked with good housing, robust Information and Communication Technology (ICT) connectivity [3], e-governance, sustainable environment, safety and security, health, and education.

Traditional infrastructure has been important in major cities in collaborating diverse activities [4]. However, integrating technology-enabled infrastructure into facilities and services would spur economic growth and create a new ecosystem. In this light, integrating technology into smart cities is critical to their progress. City video surveillance structures, central managed public utilities, integrated safety and communication, clean water management facilities, clean air, and water and electric-driven public transportation are all important considerations in the construction of smart cities.

These include waste management and recycling facilities, ample natural management areas, smart garbage disposal and composting, rainwater

DOI: 10.1201/9781003289418-3

harvesting, soil conservation, and wind and hydro power systems. The smart city is made up of this infrastructure [5].

3.2.1 Sub-Division of Smart City Infrastructure

Works done by various authors incorporate various components that fully make up a smart city. These facilities can be grouped into various sections. In the planning process, we take a look at four sections.

1. *Living space of citizens*: Forming a large part of the smart city's overall landscape, where citizens reside and perform their activities, is significant in providing a more efficient infrastructural arrangement for all smart cities' components. This consists of private living facilities and paces for diverse activities. Examples of these include houses, and vehicles.

2. *Collection of economic activities*: It is essential to look at all the financial activities going on in a smart city and develop infrastructure catered toward establishing the most appropriate facilities for business and industry. Other activities that draw economic profit for cities must also be considered. We must consider activities such as culture, sports, healthcare, and trading facilities to provide financial gain for the city.

3. *Non-physical social infrastructure*: These refer to all kinds of supportive service-oriented infrastructure supporting citizens' activities in the smart city.

4. *Physical city infrastructure*: These include the facilities necessary for supplying basic life needs in the smart city such as water and energy facilities, road, bridge, and parks.

Infrastructure [6] found in a smart city are based on its business function and may share particular identical contribution in terms of their contribution to the flow of resources in the city. It is thus essential to provide infrastructure at a city scale that incorporates technologies while at the same time performing specialized functions within the smart city. As such, there must be a way [7] for these ICT to be integrated with the infrastructure [8]. In related works, authors in [8] grouped these into the following:

1. *U-city/smart city infra*: These are city infrastructure [9] amenities that support smart city services.

2. *Smart city control center*: This is the city control center for collecting, processing, delivering, and using data across the smart city.

3. *Smart city service operator*: The smart city service provider uses data from the smart city control center.

4. *Micro infra*: This refers to infrastructure erected in a tiny living or economic realm.

3.3 Interoperability

Interoperability [10] among these specialized services is also crucial. Interoperability of all systems will be quite relevant as a foundation for information that will be built on top of the physical structure, creating an easy flow of all necessary information needed for the smart city's smooth running. However, it is quite challenging to develop an architectural or functional model of a smart city that incorporates all the sectors in a smart city. A general framework that does provide adequate consideration for all sectors is essential for the smooth running of operation all across a smart city. Some works have provided solutions, with several smart city models. Ref. [10] provides an extensive view of a smart city as a collection of infrastructures and shows the value flows within this city. The aim is to identify value flows inside a smart city and design an ICT-based system to propel innovation in streamlining information flow within the smart city. The smart city, by itself, will serve as an information infrastructure as seen in Figure 3.1.

3.3.1 Interoperability Issues

Currently, problems arising from modeling and designing smart cities are challenging to combine all the multiple subsystems and businesses

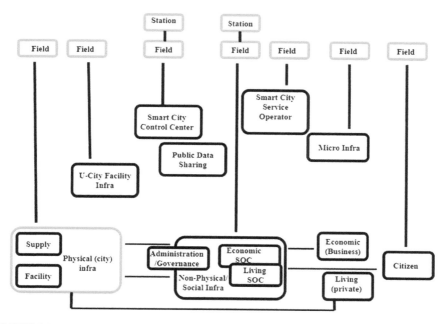

FIGURE 3.1
Interoperability between connected infrastructure in a smart city.

emerging and put them into a single infrastructure. Although we have an excellent infrastructure, what we currently have is not federated enough to ensure a synergy from one component affecting another. In normal circumstances, one infrastructure installs facilities for its smart service device. Simultaneously, other parts of the city domains develop a new business of information processing to provide smart services out of their traditional functions. This shows that these domains and components are becoming similar. There is, therefore, a need to ensure that they still work with each other. Infrastructure developed must consider these heterogeneity factors that affect city services, business domains, and stakeholders and design well-sufficient models [11]. Trying to develop a well-adaptive model is very challenging since cities are complex and infrastructure grows quickly.

3.3.2 Smart City Infrastructural Services

Since we intend to incorporate all these components and domains that make up a smart city together, there is the need to bring various services as seen in Figure 3.2 to ensure that these smart cities become much interoperable in future works and possibly become easier to connect.

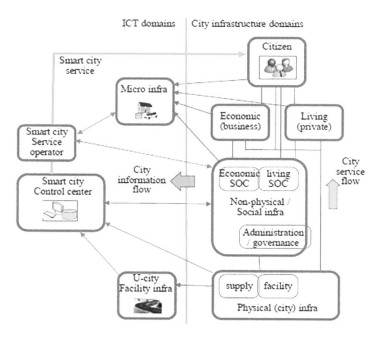

FIGURE 3.2
Proposed framework for interconnected smart city components.

Although currently, these services exist, there is still heterogenous inter-working at a low level, which makes it difficult for aggregation to occur. Normally for facilities, such as smart city grids, which are intended to generate energy, the distribution can contribute immensely to various smart city points. However, this heterogenous underworking makes it a challenge [12]. The need for other approaches such as creating interworking at a higher level (connecting various market or enterprise platforms) must likewise be considered.

3.4 Infrastructural Design

In this section, we examine the necessary architectural and engineering designs for smart cities. We examine the alterations required to create a smart-city-friendly construction system. We can conceive of features of a smart city such as multi-mode transportation; video surveillance; public utilities; integrated safety and communication domains; clean air, soil, and water; and well-adapted smart transportation. This can be seen in Figure 3.3. Assistive technologies such as solar panels and LED lighting must be incorporated into the building. Ample plantation and free area are required for rainwater harvesting and soil conservation centers, wind and hydro power facilities, as shown in Table 3.1. These must be included in the smart city design plans.

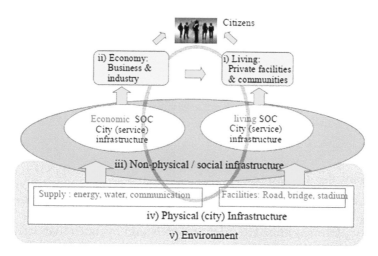

FIGURE 3.3
Infrastructure and service provisioning systems for smart city.

TABLE 3.1

Smart City Resources and Contributing Sources

Smart City Resources	Contributing Sources
Adequate water supply	Rainwater harvesting
Adequate electric supply	Green building
Effective mobility	Highly smart transport networks
Affordable housing	Low cost
IT connectivity	Decentralized data network services
Safety	Blockchain-based CCTV grids

3.4.1 Smart Homes

Homes form the essential part of smart cities and thus form the major core from which all other elements are interconnected. From the water, energy health, security, and mobility, homes help sustain the lives of occupants in a smart city. One crucial aspect of infrastructure for smart cities has to deal with smart homes. Smart homes are quite central to improvement. As such, in the development of smart city infrastructure, there is a need to plan effectively so as to provide effective building codes that will help to complement components in a smart city. Connecting other infrastructure to the smart homes is an issue of greater challenge, and for other resources such as water supply and energy, the planning of buildings must be done in such a way so that it minimizes power consumption ensuring water conservation and managing waste without any pollution. Green building technologies for standalone houses and apartments have been thus recommended [13].

3.4.2 Waste Management

In managing waste and recycling, segregation of waste at apartments and urban centers is required. Waste management centers must be created all across the smart city that allows for speedy collection of waste contents from offices and apartments. Large sewage plants with highly sensitive technologies must be built into these offices to ensure effective monitoring of waste.

3.4.3 Rainwater Distribution

The building's network must be effectively planned out and essentially details to consider the effective distribution of rainwater across the city. Smart city planners must consider channels and avenues that affect waterways and must place restrictions on the emergence of any infrastructure that may hamper the effective distribution of water across the waterways. Rainwater harvesting plants must also be created along these waterways to ensure that rainwater is collected to power the smart microgrids in the provision of electricity as well as distribute the water all across the smart city. Building regulators must also be set up to monitor such facilities' emergence and impose sanctions such as fines that can be transparently monitored on the blockchain. This framework is done if the setup provides underground water tables for river or water-dependent smart cities.

3.5 A Blockchain-Enabled Smart City Infrastructure

In this work, however, we introduce a [10] blockchain-enabled smart city that ensures the effective distribution of resources in values on the smart city network in a decentralized and verifiable manner. Participants all across the network are easily identifiable and are able to interoperate with each other to produce the best synergy needed for the operation of the smart city.

The basic goal of a blockchain-based smart city is to add a blockchain layer to the network [14], assisting value flows within a city system. However, the blockchain provides an efficient distribution channel for information across all systems. The blockchain's transparency and immutability allow for more verifiable verification of activities across the smart city. The blockchain [15] potentially provides an efficient ecology for these technologies. Water, electricity, sanitation, solid waste management, urban mobility, and public transport are linked with good housing, robust IT connectivity, e-governance, sustainable environment, safety and security, health, and education.

3.5.1 Blockchain-Enabled Infrastructure-Focused Approach

In this work, we consider a blockchain-enabled infrastructure-focused approach in designing the infrastructure necessary for a smart city.

This infrastructure is [16] combined with existing technologies such as IoT, Cloud, big data, and 5G mobile. Here ICT-based infrastructure is deployed as seen in Figure 3.4 as a means of realizing the goals of a smart city.

For each infrastructure domain with a specified function in the smart city, any actor within the domain can be easily identified based on distributed identity systems in place all over a blockchain-controlled smart city [17]. This transparent and self-verifying distributed mechanism that the blockchain provides makes it easy to know participants making contributions all over the smart city's decentralized network and, as such, participate cohesively in the effective distribution of resources.

This is more secure since a traditional infrastructure-focused approach in designing a smart city is vulnerable to malicious attacks and also prone to low underutilization due to its separation between the smart city citizens and its infrastructural platforms. Also, due to its lack of decentralization,

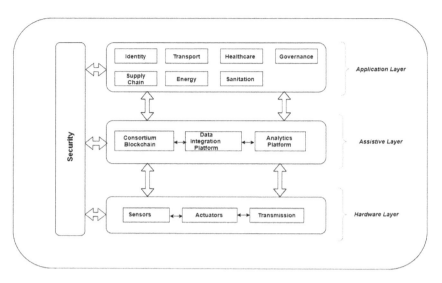

FIGURE 3.4
Operating system design underlying smart city infrastructure.

it is vulnerable to many forms of attack. Although many works have been done to make these traditional infrastructural approaches more visible and interactive with participants in the smart city, it remains a challenge. The blockchain also provides another layer of transparency and interactivity. The transparent nature of the blockchain also provides a means for participation since these citizens can be reflected as participants on the network and actions performed are recorded on the blockchain as well. These actions are easily visible by all participants on the blockchain network and can be easily seen and transactions are easily recorded. Transactions that occur within this infrastructure can also be easily observed, and this offers an excellent means of visibility and transparency that makes a blockchain-enabled infrastructure easily interactable. Table 3.2 depicts the various engineering inputs necessary to achieve a smart home based on blockchain.

3.5.2 Blockchain-Supported Smart City Platforms

Blockchain-supported platforms as seen in Figure 3.5 are created all across the distribution centers for the citizens' activities in the smart cities. These platforms must be built to run financial, sports, entertainment, and educational services where a lot of transaction takes place among various kinds of people. Payment of transactions and fees can be monitored and controlled using these platforms distributed all across the city. These platforms can be accessed through devices powered by advanced technologies such as 5G and other [18] Cloud services. This provides convenience for citizens in the

TABLE 3.2

Engineering Input Necessary for a Blockchain-Based Smart Home

Materials to Build	Inclusion Rules
Green energy building materials	For facilities like apartments and residence buildings, it is important to build these with materials conducive to green energy. Roofs must be made from green energy materials
Solar energy and LED lighting	Portion of a roof can be used for installing solar facilities and space for street lighting
Ventilation	There should be adequate planning for ventilation. Buildings must have automated openings for natural ventilation
Waste management	Waste management infrastructure should be built, and planned spaces must be provided to allocate space for waste produced in smart city
Internet connection	Buildings must have Internet-connected tools such as hubs, routers, and other IoT devices to help ensure processes within apartments and make sure homes are kept convenient.
Video surveillance	CCTV must be installed all across public streets and parks. Surveillance cameras must also be placed in homes
Blockchain-based electricity microgrids	All homes must be connected to the microgrids. Transactions and usage information can be traced using the blockchain. Also, the information provided on the microgrid can provide a lot of information to enable problem resolution on the microgrids

smart city to be able to transact effectively without having to wait for inter-mediaries. Payment of taxes, school fees, and other payments can also be facilitated easily, and thus there is the speed in moving payments from one end of the smart city to the other. This will reduce redundancy in the smart city network.

3.5.3 Blockchain-Controlled Smart City Infrastructural Regulations

With escrows, fairness can be reached across the smart city since members can collaborate effectively to build infrastructure in these smart cities. Escrows that are supported by smart contracts on the blockchain will thus ensure that misallocations are avoided in the developmental projects across a smart city and promote the rapid growth of infrastructure since all activities [19] can be effectively monitored using the blockchain. Since smart contracts provide secured access to all activities and regulate all transactions effectively by making immediate enforcement, infrastructure guidelines will be strictly adhered to and obeyed without the need for any trusted third-party regulation or organizational enforcement. The smart contract developed on the blockchain will also ensure a strict enforcement of all rules and regulations across the smart city. We develop a novel architecture from which the blockchain serves as a core entity in the smart-city-enabled network.

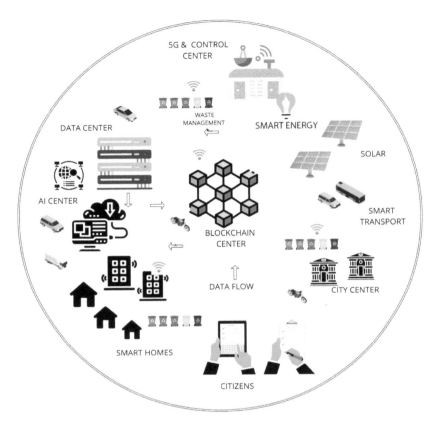

FIGURE 3.5
Proposed blockchain-based smart city framework.

3.5.4 Blockchain Unit

This is responsible for monitoring data flow of information all across various data processing centers in the smart city. The blockchain unit coordinates all smart city activities by providing an interface for interacting with data processing centers and infrastructure. Visibility is also provided for transactions that take place within the nodes found in a smart city and, at the same time, sets up an identity with which citizens in a smart city can be able to interact.

3.5.5 Analytic Platform

This serves as a decision-making center for all data and value in-flows all across the smart city. Inside an artificial intelligence unit, analysis is made of the day-to-day transaction that takes place in a smart city. The daily processes that take place among all data channels are all monitored and decisions are made about them from the Artificial Intelligence Center.

3.5.6 Smart City Control Center

This serves as the central point of the smart city. This coordinates all activities from transportation systems to health infrastructure and housing facilities. Distribution of information is also set up here. This serves as the smart city's central point and likewise provides a means of controlling all centers across it.

3.5.7 IoT Connection

This deals with the underlying network that underpins the main infrastructure of the smart city. This includes the major networks, IoT devices as well as telecommunication networks that form the vital technological components of the smart city. This is where the major protocols from which the infrastructure operates are developed.

3.5.8 Data Integration Platform

This serves as a major store of data for all activities that take place in the smart city. Information flow arrives at the data center from which they are distributed to the various units for analysis and other functional processes. The data center is the collating point of smart city and ensures that data integrity is secured.

3.5.9 Data Type Aggregation Center and Transmission

This provides the necessary network infrastructure that ensures the conversion of aggregated data into standardized data types. This is sent to the various parts of the network to support devices and components within the smart city. It checks information such as the bandwidth of networks, connected devices on all networks and a data type matching register that matches the data and compares them to pre-loaded stored data formats.

3.5.10 Infrastructure Security Consideration

An operating system's functions must be protected from other smart city components.

So that processes that seek to access files, memory, CPU, or other hardware resources are properly authorized by the operating system. A process's address space on the blockchain can be verified by memory addressing hardware. So no process can control the CPU without relinquishing it. Finally, no process is allowed to execute its own I/O, which helps maintain the integrity of the smart city network's peripheral devices.

References

[1] Ghosh, S. (2018, March). Smart homes: architectural and engineering design imperatives for smart city building codes. In 2018 Technologies for Smart-City Energy Security and Power (ICSESP) (pp. 1–4). IEEE.

[2] Arora, S. (2008). National eID card schemes: a European overview. Information Security Technical Report, 13, 46–53

[3] Ahn, J. Y., Lee, J. S., Kim, H. J., & Hwang, D. J. (2016, July). Smart city interoperability framework based on city infrastructure model and service prioritization. In 2016 Eighth International Conference on Ubiquitous and Future Networks (ICUFN) (pp. 337–342). IEEE.

[4] Lv, Z., Hu, B., & Lv H. (2020, March). Infrastructure monitoring and operation for smart cities based on IoT system. IEEE Transactions on Industrial Informatics, 16(3), 1957–1962. doi: 10.1109/TII.2019.2913535.

[5] Albino, V., Berardi, U., & Dangelico, R. M. (2015). Smart cities: definitions, dimensions, performance, and initiatives. Journal of Urban Technology, 22(1), 1–19.

[6] Al-Hader, M., & Rodzi, A. (2009). The smart city infrastructure development & monitoring. Theoretical and Empirical Researches in Urban Management, 4(2[11]), 87–94.

[7] Chourabi, H., Nam, T., Walker, S., Gil-Garcia, J. R., Mellouli, S., Nahon, K., … & Scholl, H. J. (2012, January). Understanding smart cities: an integrative framework. In 2012 45th Hawaii International Conference on System Sciences (pp. 2289–2297). IEEE.

[8] Serrano, W. (2018). Digital systems in smart city and infrastructure: digital as a service. Smart Cities, 1(1), 134–154.

[9] Gascó-Hernandez, M. (2018). Building a smart city: lessons from Barcelona. Communications of the ACM, 61(4), 50–57.

[10] Li, S. (2018, August). Application of blockchain technology in smart city infrastructure. In 2018 IEEE International Conference on Smart Internet of Things (SmartIoT) (pp. 276–2766). IEEE.

[11] Adtell Integration (2018), "Smart Cities Infrastructure – Adtell Integration – Total Communication Solution," adtellintegration.com [Online]. Available: http://adtellintegration.com/smart-cities-infrastructure/. [Accessed: 16 April 2018].

[12] Wang, P., Ali, A., & Kelly, W. (2015, August). Data security and threat modeling for smart city infrastructure. In 2015 International Conference on Cyber Security of Smart Cities, Industrial Control System and Communications (SSIC) (pp. 1–6). IEEE.

[13] Getty Images (2018), "Smart City Concept and Internet of Things Stock Vector Art; More Images of City 510687292 iStock," www.istockphoto.com [Online]. Available: https://www.istockphoto.com/vector/smart-city-concept-and-internet-of-things-gm510687292-86357031. [Accessed: 16 April 2018].

[14] Hao, L., Lei, X., Yan, Z., & ChunLi, Y. (2012, June). The application and implementation research of smart city in China. In 2012 International Conference on System Science and Engineering (ICSSE) (pp. 288–292). IEEE.

[15] Hmood, R., Katib, S. S. I., & Chlamtac, I. (2020). Smart Infrastructure and Applications. Springer International Publishing.

[16] Nam, T., & Pardo, T. A. (2011, September). Smart city as urban innovation: focusing on management, policy, and context. In Proceedings of the 5th International Conference on Theory and Practice of Electronic Governance (pp. 185–194). New York: Association for Computing Machinery.

[17] Sarwat, A. I., Sundararajan, A., Parvez, I., Moghaddami, M., & Moghadasi, A. (2018). Toward a smart city of interdependent critical infrastructure networks. In Sustainable Interdependent Networks (pp. 21–45). Springer, Cham.

[18] Su, K., Li, J., & Fu, H. (2011, September). Smart city and the applications. In 2011 International Conference on Electronics, Communications and Control (ICECC) (pp. 1028–1031). IEEE.

[19] Schatten, M. (2014). Smart residential buildings as learning agent organizations in the internet of things. Business Systems Research Journal, 5(1), 34–46.

4

Identities

4.1 Introduction

Nationality is a cultural and linguistic group entitled to self-government [1]. An identity is a set of characters that uniquely identifies an individual or a group. A national identity is a government-provided unique identity for its citizens. Indonesia has 29 federal identity cards issued by 24 institutions. These include passport, driver's license García et al. [2], tax identification number, and insurance. Each identification is unique to the agency issuing it. Demographic data proves to be ineffective due to lack of collaboration between government entities and associated information framework. To register, citizens must go to several organizations with varying techniques. Replication, redundancy, and wasteful population data and information are inevitable. It is also difficult to communicate important information because each government agency has its own database.

Several countries have adopted digital identity policies. Because digital identification involves sensitive information, it is used to confirm or verify a person's identity [3]. Governments store this data in centralized databases. Because hackers may quickly enter into such systems and reach their harmful aims [4], personal data housed in centralized databases becomes a key target for data breaches. National governments, with the help of development partners, are working hard to enhance public administration, plan development, enforce the law, and fight corruption. Lots of governments are planning national digital identity programs. These government-coordinated schemes aim to equip citizens with a single digital identity.

In Indonesia, a single digital identity (SDI) incorporates data from several government and private sectors. The business sector's involvement in digital identity shows the growth of new public-private partnerships to build and strengthen the proof framework in emerging nations. The Ghanaian government recently introduced a system for registering SIM cards. User identity is required to access advanced administrations and communications. Interfacing a digital or biometric identity with a SIM card raises questions of belief and ownership of personal information. In order to ensure that advanced digital identification frameworks are compelling,

DOI: 10.1201/9781003289418-4

safe, comprehensive, and reliable, both public and private partners must collaborate.

The global identity framework disseminates through a decentralized instrument, and blockchain serves as an open repository service for each transaction. An identity can be used for a person and shared on the blockchain network. Clients may control how data is modified and overhauled using blockchain, which is a distributed record. Blockchain innovation keeps data at distinct nodes, and data is added when the nodes agree. The fact that previous data cannot be destroyed when new entries are added makes it straightforward to follow transaction history. The use of blockchain signifies the next step in e-government. It can help save expenses and complexity, share trustworthy forms, broaden review trial discoveries, and ensure reliability. Blockchain technology is characterized by simplicity and faith. Regardless, there are situations when information security is required in business transactions.

4.2 Digital Identities

Digital identity is a simple idea to grasp. Digital is any personal data found online that can help identify you. For example, photos shared on social media, postings published or comments made, online bank account, search history, etc. Despite this, we are generally able to give our online personas solidity and data. Whether physical or digital, identity is an amalgamation of personal characteristics that allows a person to be identified and interact with others. Coherent reflections of identification show that group transactions are required. Then identities are created to facilitate trades. Individuals, legal substances, resources, forms, or any substance whose actions affect the system can be identified. We associate identities with enterprises, organizations, and trusts. Resources include tangibles like automobiles and buildings, as well as intangibles like licenses and data. In general, identities connect to people. Each entity's identity is based on three fundamental attributes.

Inherent attributes are traits that an entity has that are not determined by external variables or substances. Age, date of birth, fingerprint and height are all inherent qualities of persons. Names are given to attributes gathered through time.

Accumulated attributes regularly change over the entity's life. Typical examples are the person's employment history and Facebook friends list.

Assigned attributes are properties that organizations offer to entities. The traits in question may be disputed due to a person's affiliations with the organization. Individual examples include international id, advanced login IDs, passwords, and phone numbers.

4.3 Digital Identity Life Cycle

There are three essential stages of digital characters life cycle.

1. *Registration*: The first step in creating a digital identity is to register. Personal information (e.g., title, date of birth, address, sex), biometrics (e.g., unique mark, iris filter) and other associated attributes are captured and recorded. Identity is evaluated for presence, uniqueness and links.

2. *Issuing of credentials*: Typically, ID issuers give documents (birth certificates) or eIDs (ePassports). Other electronic accreditations include smart cards and Cloud-based ID.

3. *Authentication*: Authentication is the process of verifying and approving the record's ownership. The client must be authorized using one or more factors that fit into one of four categories: being, having, knowing, or doing.

4.4 Life Cycle Management

Life cycle management facilitates updating the status and content of digital identities. The client should occasionally update their marital status, age, and profession. Alternatively, the issuer of the identity may reject the digital identity for security grounds and terminate it in the event of death.

4.5 Identity Evolution

Identity has gone through four stages: centralized, federated, user-centric, and self-sovereign.

1. *Central identity*: Control by a single expert or hierarchy. For example, while issuing a national ID card, the government oversees and stores all personal data in one central database.

2. *Federated identity*: It enables sharing of IDs and attributes among organizations in a defined circle of trust, such as citizens using national identity providers.

3. *User-centric identity*: By decentralizing identity, clients can have greater protection and control over their personal data.

4. *Self-sovereign identity*: An individual has sole control over their account and personal information. They can be independent of central authorities if they are sovereign.

They can validate and communicate information on their devices. They also have full control over how their data is stored and used.

4.6 How Does the Digital Identity Work?

4.6.1 For Companies

Companies routinely acquire sensitive client information and store it alongside less-sensitive transactional data [5]. It also shifts the industry attention to corporate IT obligations, which raises additional commerce risks. When these data are locked up in secret data vaults, they lose their value in generating product improvements and true customer comprehension. Many efforts seek costly and risky enterprises to achieve the ideal balance between information security and commercial needs.

4.6.2 For IoT Devices

Around 7 billion devices are web-enabled. This number is expected to double by 2020 to 10 billion. Unlike the early Web, which only consisted of trusted institutions [5], most IoT inventions lack adequate identity and access control capabilities. Sensors, screens, and devices must be recognized by IoT gadgets and objects to secure sensitive and non-sensitive data. Leading IT vendors are now offering IoT management frameworks to address these issues. For example, a single firm may have tens of thousands of IoT devices compared to a few dozen or so traditional servers and user devices. With so many devices, it's easy to mismatch guidelines. Large-scale IoT hacking is becoming a hot topic at top IT security conferences.

4.6.3 For Individuals

Working societies and economies require identity. Existing social structures and global marketplaces thrive when we can legitimately identify ourselves and our possessions. Identity is a set of statements about an individual, location, or thing. The personal information includes first and last name, birthdate, nationality, and a few forms of national identification such as international id number, SSN, driving license, etc. Centralized authorities (governments) issue and store these data points (central government servers).

A human's access to physical forms of identity varies. Around 1.1 billion people worldwide lack a mechanism to assert ownership of their identity [6]. This leaves one-seventh of the world's population defenseless, unable to vote, own property, open a bank account, or find job. Inability to obtain identification documents jeopardizes access to the financial system and limits opportunities. Only a few citizens have complete ownership and control over their identities. Unconsciously, they lose the esteem that their data creates [7]. Data-holding companies are frequently hacked, forcing end-users to live with extortion moderation. When a social security number is lost, there is little to no response.

4.7 The Need for Blockchain-Based Identities

Blockchain can be utilized to overcome modern identification difficulties. Modern identity issues include inaccessibility, data security, and fraud.

4.7.1 Inaccessibility

Around 1.1 billion people lack proof of identity, with 45% being among the world's poorest 20% [8]. Costs, availability, and a fundamental lack of information about individual identity keep nearly a billion individuals outside of standard identification systems. The lack of a physical identity prevents access to many government services and educational opportunities. Having a name is important for accessing the existing monetary structure. However, 60% of the unbanked claim mobile phones, paving the way for blockchain-based mobile identity solutions that better suit vulnerable citizens [8].

4.7.2 Data Insecurity

Currently, our most vital identification data is stored in centralized government systems supported by legacy software with several points of failure. Huge centralized frameworks hosting millions of customer accounts attract hackers. According to a recent survey, 97% of all data breaches in 2018 involved personally identifiable information. Despite legislative and corporate efforts to improve cybersecurity, 2.8 billion consumer records were exposed in 2018 [9].

4.7.3 Fraudulent Identities

The user's digital identity landscape experience is also highly fragmented.

Usernames are used to juggle many identities on various websites. There is no standard mechanism to use data created by one system on another.

Moreover, the weak link between digital and offline identities makes faking identities straightforward. Phony engagement is a phenomenon that can aid in extortion, increase numbers, and lose income. This susceptibility in society facilitates the production [10] and spread of disasters like "false news," which threatens democracy.

4.8 Decentralized Digital Identities

Clients can create and manage digital identities using decentralized identifiers, identity management, and embedded encryption.

4.8.1 The Creation of a Digital Identity

Clients register a decentralized identification with a self-sovereign identity and data system. During this time, the client creates secret and public keys [11]. If keys are compromised or rotated for security reasons, public keys can be stored on-chain. To make the system scalable and meet with privacy rules, extra information linked with a decentralized identity, such as attestations, can be safeguarded on-chain.

4.8.2 Decentralized Identifier

Individuals, institutions, assets, and so on may all have decentralized identifiers. A secret key protects each DID. Only the secret key owner can prove ownership of a given identity [12]. A person can have multiple DIDs, limiting their ability to do multiple workouts. A person may have one DID for a gaming site and another for a credit reporting platform [13]. Each DID often has many attestations from other DIDs that confirm specific DID traits (e.g., hometown, height, recognitions, age). This allows DID owners to store their own credentials rather than relying on a single profile provider (e.g., Google and LinkedIn). Depending on the context and intended usage, the owner of the material can also link non-attested information like social media posts to DIDs.

4.8.3 Security of Decentralized Identities

Cryptography is used to safeguard decentralized identities. Secret keys are solely known by their owners in cryptography, whereas public keys are widely spread. This pairing serves two purposes [14]. First, the public key verifies that the message was sent by the matched private key holder. The second is encryption, where only the possessor of the matching private key may decipher the communication.

4.8.4 The Use of Decentralized Identities

Clients can exhibit their confirmed identification in the form of a QR code to access specific administrations once linked with a decentralized identity. The service provider verifies the identification by validating the attestation's control or ownership. The client signs the presentation with the DID's private key [15]. Then access is granted.

4.9 The Use of Blockchain in Managing Identities

Decentralized and digital identities have many applications. Some notable use cases include self-sovereign identity, data monetization, and data portability.

4.9.1 Self-Sovereign Identity

Self-sovereign identity is the idea that individuals and businesses can store their claim identity information on their devices and choose which information to share with validators. No nation-state, organization, or global organization is required to create these identities [16].

4.9.2 Data Monetization

Blockchain-based self-sovereign identities and decentralized models provide clients control over information monetization. Data monetization refers to monetizing personal data for financial gain. Data has value on its own, but knowledge derived from personally identifiable data increases the value significantly. Every day, 4.39 billion web users create quintillion bytes of data. By 2022, almost 60% of global GDP will be digital, increasing the value of individual data [17]. Our online data is currently ethereal, invisible, and sophisticated. Self-sovereign identity allows you to link your online data to your decentralized identify. People can then monetize their data by leasing it to AI training algorithms or selling it to sponsors. Clients could also choose to keep their data hidden from groups or governments.

4.9.3 Data Portability

The General Data Protection Regulation (EU GDPR) gives clients the right to data management, which means they can have their data directly transferred from one controller to another, if technically practicable. This right can improve customer experiences by reducing the requirement to authenticate identity across administrations and platforms [18]. With DIDs and undeniable accreditations, moving identities from one target system to another is simple. While information portability reduces client friction, expediting the sign-up process promotes client selection. Clients can quickly reverify themselves using DID data portability, meeting administrative Know Your Client (KYC) requirements. As a result of eliminating the laborious identification valida-tion process where generally a part of records must be given and examined, the financial division can save money.

4.10 People

Buildings are "machines for living," and cities are collections of buildings and accompanying infrastructure. Future cities will aim to improve the effi-ciency of machinery by embracing cutting-edge innovation [19]. Cities can be viewed as self-contained organisms having both mechanical and human components. Humans are essential for cities to become economic, socially useful, and environmentally sensitive places to work, share, and live. Despite significant increases in the level and intelligence of mechanization in well-designed future cities, each benefit and action ultimately impacts individuals [20]. That should be the focus of city planning and operation. Aside from that, people who live in cities tend to do so in non-exclusive areas, which might affect their relationships with the city and one other.

4.11 Conclusion

Demands for privacy and openness are everywhere nowadays. Because of its security and transparency, blockchain will lead today's globe in progressive innovation. Identity is a vital component of our everyday lives as citizens, and digital identity will help to simplify our lives. The use of blockchain in digital identification is essential to maintain it secure. So this chapter focused on the role digital identities play in smart city administration. The benefits of blockchain technology for digital IDs in smart cities were examined.

References

[1] S. Arora, "National eID card schemes: A European overview," Inf. Secure. Tech. Rep., vol. 13, pp. 46–53, 2008.

[2] C. G. García, D. Meana-Llorián, B. C. P. G-Bustelo, J. M. C. Lovelle, and N. Garcia-Fernandez, "Midgar: Detection of people through computer vision in the Internet of Things scenarios to improve the security in Smart Cities, Smart Towns, and Smart Homes," Future Gener. Comput. Syst., vol. 76, pp. 301–313, 2017.

[3] Adtell Integration, "Smart Cities Infrastructure – Adtell Integration – Total Communication Solution," adtellintegration.com, 2018. [Online]. Available: http://adtellintegration.com/smart-cities-infrastructure/. [Accessed: 16 April 2018].

[4] V. Albino, U. Berardi, and R. M. Dangelico, "Smart cities: Definitions, dimensions, performance, and initiatives," J. Urban Technol., vol. 22, no. 1, pp. 1–19, 2015.

[5] L. G. Anthopoulos and C. G. Reddick, "Smart city and smart government," in Proceedings of the 25th International Conference Companion on World Wide Web – WWW '16 Companion, 2016, pp. 351–355.

[6] J. Gao, K. O. Asamoah, E. B. Sifah, A. Smahi, Q. Xia, H. Xia, ... and G. Dong, "GridMonitoring: Secured sovereign blockchain based monitoring on smart grid," IEEE Access, vol. 6, pp. 9917–9925, 2018.

[7] S. Borlase, editor, Smart grids: infrastructure, technology, and solutions. CRC Press, 2017.

[8] M. Angelidou, "Smart cities: A conjuncture of four forces," Cities, vol. 47, pp. 95–106, 2015.

[9] S. Zygiaris, "Smart city reference model: Assisting planners to conceptualize the building of smart city innovation ecosystems," J. Knowl. Econ., vol. 4, no. 2, pp. 217–231, 2013.

[10] P. Lombardi, S. Giordano, H. Farouh, and W. Yousef, "Modelling the smart city performance," Innovation: Eur. J. Soc. Sci. Res., vol. 25, no. 2, 137–149, 2012.

[11] T. Bakici, E. Almirall, and J. Wareham, "A smart city initiative: The case of Barcelona," J. Knowl. Econ., vol. 4, no. 2, pp. 135–148, 2013.

[12] J. R. Gil-Garcia, T. A. Pardo, and T. Nam, "What makes a city smart? Identifying core components and proposing an integrative and comprehensive conceptualization," Inf. Polity, vol. 20, no. 1, pp. 61–87, 2015.

[13] R. Rivera, J. G. Robledo, V. M. Larios, and J. M. Avalos, "How digital identity on blockchain can contribute in a smart city environment," in 2017 International Smart Cities Conference, ISC2 2017, 2017.

[14] I. Thomas and C. Meinel, "An identity provider to manage reliable digital identities for SOA and the web," in Proc. 9th Symp. Identity Trust Internet – IDTRUST '10, 2010.

[15] G. Ben Ayed and S. Ghernaouti-Hélie, "Digital identity management within networked information systems: From vertical silos view into horizontal user-supremacy processes management," in Proceedings – 2011 International Conference on Network-Based Information Systems, NBiS 2011, 2011.

[16] Q. Xia, E. B. Sifah, K. O. Asamoah, J. Gao, X. Du, and M. Guizani, "MeDShare: Trust-less medical data sharing among cloud service providers via blockchain," IEEE Access, vol. 5, pp. 14757–14767, 2017.

[17] E. B. Sifah, A. Shahzad, S. Amofa, S. Abla, K. O. B. O. Agyekum, Q. Xia, and J. Gao, "Blockchain based monitoring on trustless supply chain processes," in Proc. IEEE Int. Conf. Internet Things (iThings), IEEE Green Comput. Commun. (GreenCom), IEEE Cyber Phys. Social Comput (CPSCom), IEEE Smart Data (SmartData), July, 2018, pp. 1189–1195.

[18] J. Yong, "Digital identity design and privacy preservation for e-learning," in Proceedings of the 2007 11th International Conference on Computer Supported Cooperative Work in Design, CSCWD, 2007.

[19] E. F. Lambin, H. J. Geist, and E. Lepers, "Dynamics of landscape and land cover change in tropical regions," Annu. Rev. Environ. Resour., vol. 28, no. 1, pp. 205–241, 2003.

[20] X. Luo, Y. Ren, J. Hu, Q. Wu, and J. Lou, "Privacy-preserving identity-based file sharing in smart city," Pers. Ubiquitous Comput., vol. 21, no 5, pp. 923–936, 2017.

5

Supply Chain

5.1 Introduction

In a smart city, there is a need for a well-coordinated system that is transparent, verifiable, and easily networked. For a smart city interconnected worldwide globally, many goods of various kinds are shipped from one part of the world to the other in large quantities and between very complex paths. These paths are intertwined between countries, cities, and towns. As such, detecting information and providing verifiable proofs of activities remains a great challenge in the supply chain. Every day goods are shipped in large quantities across the globe to be burned to produce power in smart cities. Wheat and rice produced in various parts of the world are shipped to countries all over the world. These routes involve multiple parties who play a diverse role. Parties include consumers purchasing goods at warehouse centers, retailers, automated transport systems, and various electronic sales point.

The effective coordination of each party affects the supply chain altogether. An efficient mechanism to ensure proofs of activities on the supply chain ensures trust among parties who may distrust each other on the supply chain. This will ensure that the agreed-upon result is achieved and provide an avenue for members of the chain to participate efficiently. Members such as farmers of the goods will fully trust other parties for the safe arrival of their goods without being tampered with or changed for no reason on the supply chain. Also certifying the originality of goods and its quality ensures an entire population's health and sustainability in a country. There is, therefore, the need to ensure that proof of all the activities on the supply chain is effectively done otherwise a large number of unverifiable [1] actions could result in a huge number of losses for all parties involved in the supply chain network. Proving activities will also ensure that the goods end up at its required destination and provides trust [2] among coordinating parties.

However, proving activities on the supply chain remains a challenge due to the complexity of the supply chain. Most systems today [3] use the traditional logistic systems that track these goods and stores and provide delivery details. There is an inefficient delivery verification system when delivering goods from one point of the supply chain to another. Means of payment also

DOI: 10.1201/9781003289418-5

remain non-transparent after delivering goods that do not provide trust in payment [4]. Systems do not exist that ensures units comply with ethics and regulations on the supply chain. There also exists a lack of trust in payment. Also, due to the inability to trace goods in a decentralized manner, security is not ensured and as such security of goods remains a challenge. Again, due to the extreme need to ensure security in the supply chain, verification processes take a long-time causing delays in the arrival of goods that do cause more complications. There it shows that existing mechanisms for ensuring trust in the delivery of agricultural goods are not as efficient in handling the issues in trust and proofs on supply chain.

Since it is important to provide proofs of whatever takes place on the supply chain, there is the need for a more efficient proving mechanism, something the blockchain provides. This dissertation thus aims to ensure that activities on the supply chain are effectively proven using blockchain technology to therefore ensure the safe delivery of goods in a smart city.

We will use Game Theory (GT) to reward participation on our blockchain-based network in later sections of this effort because the supply chain relies on mutual confidence among participants to ensure its overall profitability. However, participants can act selfishly in their own interests, resulting in sub-optimal supply chain outcomes. Thus, efficient collaboration among all partners is required to ensure supply chain profitability. A supply chain profit can be maximized by using contracting methods to govern IoT devices and nodes. In our smart city concept, blockchain may be utilized to secure, trust, and benefit the supply chain. We create a smart contract that uses blockchain and GT to calculate each node's expected return and the supply chain's projected yield. It provides them with the best payoff that will satisfy all nodes based on the decisions of each individual on the supply chain. The GT models are not to force the supplier and retailer into making a choice but to provide the best options among their demands made, to make the best decision. Our smart contract automates each process while, at the same time, ensuring trust among participating nodes on the chain as seen in Figure 5.1.

5.2 Blockchain in Supply Chain Framework

On the supply chain network for smart cities, the blockchain is leveraged to provide a more effective Proof of Delivery (PoD) system for goods from entities on the supply chain such as farmer to the retailer and subsequently to the consumer. Due to security features of the blockchain to monitor goods from one end of the chain to another, security is enhanced. The solution is cycle-independent, which means that once goods have begun, they can be routed along different paths to get to the required destination. Although paths and routes are pre-specified to maintain routing efficiency, in moments of an

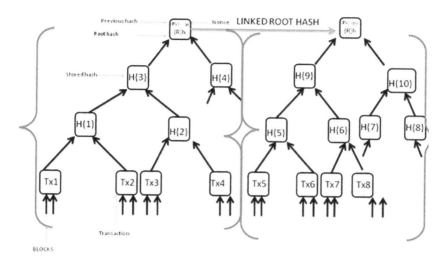

FIGURE 5.1
A Merkle tree structure for Bitcoin blockchain.

emergency such as weather conditions, a way is provided to take alternative paths while still maintaining trust in the system. As far as the last destination is not reached, entities responsible for processing goods can take an alternative route to parties in the processing phase found on the blockchain as seen in Figure 5.2. Each processing entity currently in possession of the goods specifies the next recipient of the goods that is logged into the proposed blockchain system, and thus the next recipient information is known and viewed as a trusted member. This scheme also has a cross-border advantage as movement and quality of goods can be effectively proven across different countries and even along with different distances. The conditions of goods are also monitored as slight changes to goods determine the quality of goods expected. Thus, standards of goods that are moved and brought to consumer can easily be compared and observed.

The proposed solution here presents a practical and secure way of proving goods' movement from across different endpoints and locations using blockchain technology. Participating parties known as processing entities are used. An observing party is also created to ensure that all parties involved keep to the rules and regulations involved. These entities are referred to as the PolicyChecker. Again to ensure a complete cycle and observation, another party is added to the monitoring process. This entity is known as the consumer. Finally, entities are assigned their roles on the blockchain. This creates a distributed network of entities whose role is to process these goods and maintain their standards while at the same time moving them from one end of the chain to another. A smart contract is established for all parties under each category, that is, PolicyChecker, processing entities and

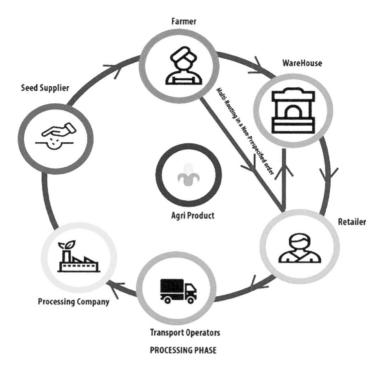

FIGURE 5.2
Processing phase diagram for proposed blockchain-enabled supply chain.

consumer. Various functions are assigned to each category with each category performing certain functions that are inherent to it. Each category can have as many individuals as possible, and blockchain addresses are assigned to every individual under these categories on the blockchain.

These categories themselves are defined as contracts that have their own addresses stored. Thus, smart contracts can interact with each other using their addresses given. Addresses of individuals in smart contracts can also interact with other smart contract addresses to ensure the smooth passage of goods and requisite services. It is important to note in this system that there is efficient monitoring of goods. All these three smart contracts must be set up. The PolicyChecker ensures that the terms and conditions are kept between processing entities and that processing entities keep to rules and regulations in order to avoid penalties and fines on the blockchain. The roles of entities as well as several terminologies used are summarized as follows:

1. *Processing entities*: These are bodies or group of individuals responsible for the systemic processing of goods. They refer to companies and bodies the responsibility of which is to cross-check and deal with the goods. This includes the suppliers, farmers, shipping companies,

customs offices, sales and merchandized purchase, processing companies, and other entities that ensure that goods state is effectively monitored and authorize transactions of goods from one end to the other. Smart contracts also regulate the actions of participants under the same name.

2. *PolicyChecker*: These ensure that rules are kept on the chain. They ensure that processing entities keep to their end of the deal after each transaction [5]. They monitor verification, and other authentication processes as processing entities move goods from one end of the chain to another. They monitor with an escrow-based policy. They store escrow fees at the beginning of each processing phase and demand the payment of this escrow before the transaction. They also punish all entities that fail to adhere to rules by taken away all escrow fees as punishment.

3. *Consumer*: This is the last end of the supply chain. The consumer can interact with processing entities in order to request for particular information.

4. *Warehouse*: Here the goods are kept in the right storage area. Inventories are also kept here. These inventories are kept by processing entities who log them onto the blockchain. To ensure that perishable goods are kept safe, the right temperature of goods is maintained at the warehouse and fully monitored to ensure proper conditions are kept.

5. *RFID*: The RFID is a set of a unique identifier for identifying goods. By combining an RFID reader and tags, the RFID of goods is read, thus providing a paperless format to accurately monitor temperature changes, uniquely identify a product and improve efficiency and accuracy.

6. *GPS*: GPS receiver is used to observe the geographical location of goods so as to indicate arrival time and where they are positioned as they move from one end of the chain to the other. This information is then logged into a decentralized or centralized database (in case of large data), and relevant parts are hashed and inputted onto the blockchain.

7. *Verifying keys*: This refers to asymmetric encryption keys used to sign goods that have arrived at one point of the chain. It makes use of blockchain's elliptic curve signature mechanism to authenticate transactions as goods are passed from one processing entity to another. There is the need to authenticate the entity at the receiving end. The processing entity upon arrival signs and verifies that all conditions are met. Through point-to-point checks, we ensure that the goods are being handed to the right party. Once this is fulfilled, he signs and receives his payment for delivery.

8. *Re-encryption keys*: This refers to the asymmetric encryption key, as seen in Figure 5.3. They provide an access-control mechanism for the processing entities. Upon delivery, the processing entity locks the information through a permission-based system. If there is a request to view the transaction, a re-encryption key is created by the processing entity to provide access to the consumer or whoever will like to view the transaction on the blockchain. The consumer or whoever requests to view transaction interacts with the processing entity using this key.

9. *Transactions*: A transaction refers to a message that is sent from one account to the other. Transactions can be sent to contracts or addresses in contracts.

10. *Gas*: Each transaction is charged with a certain amount of gas. A gas price is paid upfront whenever a transaction is set to be executed. A gas limit is set on each transaction to ensure that the computation cost does not exceed a particular number of limitations as set by the smart contract creator.

11. *Events*: Events store information or arguments as transaction logs on the blockchain's Virtual Machine (EVM). These logs are associated with a particular smart contract's address and transactions that occurred on the chain. They are registered information of transactions that have taken place on the chain permanently on the blockchain. They stay there as long as the block is accessible.

5.2.1 A Blockchain-Enabled Supply Chain Architecture for Smart Cities

Figures 5.2 and 5.3 show that from the origin of these raw materials and goods by parties such as farmers, the information of the raw materials is entered onto the blockchain [6]. Information such as temperature, quantity and weight of agri-good and other information is entered onto the blockchain. The details of the farmer are also taken as well as very sensitive information such as his blockchain address and roles he performs on the chain. This is entered into a public decentralized ledger so that everyone can be able to view. The unique identifier of the agricultural product such as the RFID and agri-good location is also collected and logged into the decentralized blockchain.

Before starting the processing phase, the blockchain-enabled system provides an alternative for the supplier who intends to begin the processing phase to agree with the smart contract defined or not. He has the choice to reject if he disagrees. The system continues by logging in the specification of the processing entity who requested the goods as well as the time frame of goods expected and certain conditions that must be met in order for the goods to be acceptable. For instance, pork must be of a specific temperature and has a certain limited amount of days before they expire. All this information is logged onto the blockchain system.

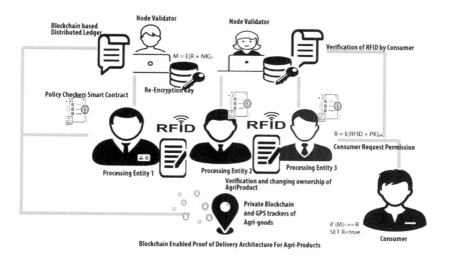

FIGURE 5.3
Blockchain-enabled goods supply chain architecture.

5.2.2 Good Processing

Before starting the processing phase, the parties must agree to the smart contract. The system continues by logging in the specification of the processing entity who requested the goods and the time frame of goods expected and certain conditions that must be met for the goods to be acceptable. All this information is logged onto the blockchain system. Importers/exporters, processing facilities, distribution centers, and stores are examples of processing entities that push blockchain data. The agricultural goods are packaged and provided with an RFID, unique identifier. Information such as nutritional components is logged and hashed on the blockchain. The processing entities associated with these particular goods are also logged unto the blockchain. Once all these are specified, the processing phase is initiated. The processing entities move the good from one recipient to the next. If they fail to obey the rules, the PolicyChecker enact a penalty through an escrow-based system. Each processing entity pays an escrow fee before beginning the process. Upon fulfilling delivery, his escrow fees are refunded, and payment made automatically on the blockchain. Upon arrival at each point, the goods are checked using the smart contract, and the next processing entity is also checked to see if it satisfies the requirement before goods are transferred. If goods are handed successfully, payment is enacted.

5.2.3 Verification

Once goods arrive at the end of the supply chain such as retailer, a request is made by the consumer to view all transaction associated with the agricultural goods. For the request, the goods' RFID is parsed unto the blockchain network to the processing entity. These are parsed along with the public key of consumer. The information is encrypted [7] and sent back to the consumer. If the message resolves to the request, the request outputs as true. The consumer can verify the agricultural goods on the blockchain by viewing all transactions.

5.2.4 Considerations

Consideration of other parties, as well as the timestamp of handing over goods, is also noted. The goods are verified as they move from one individual to the other using an elliptic curve cryptographic signature system on the Ethereum blockchain. Verification is done about whether they are the next intended recipient of the goods or not. If not certain conditions of punishment are specified for anyone who tried the malpractice of acting as the imposter. A certain amount is taken from his balance on the blockchain by the PolicyChecker. Upon satisfying conditions, the goods are moved to the next recipient, and the information is logged onto the blockchain.

The proof is shown of the quality of the agricultural goods as it moves, and the proof of their activities is recorded efficiently on the blockchain. The information remains consistent due to the consensus ability of the blockchain. Goods must arrive within the time frame specified. Each processing entity thus has a defined time frame in which to deliver the goods to the next processing entity ensuring that goods delivered arrive on time or else upon arrival, only a certain percentage of the agreed-upon payment will be issued to the party that delivered the blockchain. Again, if the specifications are unmet, only a specified amount of the agreed-upon payment will be issued to the party.

The design also takes into consideration other parties that may take part in the supply chain by logging their details as well as the timestamp of when they were handed the goods. This is hashed and stored on the blockchain system. The goods are verified as they move from one party to the other using an elliptic curve cryptographic signature system on the Ethereum blockchain. Verification is done of whether they are the next intended recipient of the goods or not. If not certain conditions of punishment are specified for anyone who tried the malpractice of acting as the imposter. A certain amount is taken from his balance on the blockchain by the PolicyChecker. Upon satisfying conditions, the goods are moved to the next recipient, and the information is logged onto the blockchain.

The quality of the goods is shown as it moves on the supply chain and their activities [8] are recorded efficiently on the blockchain. This information is logged on the blockchain unto a decentralized ledger. This ledger can be viewed by multiple parties across different points of the chain. The information remains consistent due to the consensus ability of the blockchain. Goods must arrive within the time frame specified. Thus, each processing entity has a defined time frame in which to deliver the goods to the next processing entity. This ensures that goods delivered arrive on time or else upon arrival at the end of the supply chain, only a certain percentage of the agreed-upon payment will be issued to the supplier by the blockchain system. Again, if the specifications detailed in the contract are not met, only a certain specified amount of the agreed-upon payment will be issued to the supplier.

Information from the original point to the last point on the supply chain is shown. The consumer can decide on that point whether he will like to purchase or not from the proof provided by the blockchain system. The figure describes the flow of the supply chain processing phase.

At certain lines on the supply chain such as warehouse, the information is stored on a private blockchain. Other important information such as location, temperature and other metrics are measured using sensors to detect changes in conditions. The GPS tracker is used to monitor the goods' location as it is being shipped or transported from one end of the supply chain to the other. Subsequently, the hash of information is entered into the blockchain system where it can be used as proof and cannot be tampered with.

Payment of goods after delivery and verification is also taken care of since, after delivery, there is payment immediately made on the blockchain system to the party's address that made the delivery. A receipt from the blockchain is also issued as proof of payment immediately. This would have taken days in traditional systems.

Consumer upon purchasing the goods can prove the originality and quality of the goods are bypassing the goods RFID onto the blockchain system. All the information from the original producing place to the last point on the supply chain will be shown. The consumer can make a decision from that point whether he will like to purchase or not from the proof provided by the blockchain system. Figure 5.4 describes the flow of the supply chain processing phase.

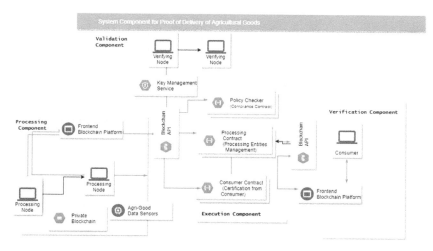

FIGURE 5.4
System component of the proof of delivery architecture.

5.2.5 Operational Requirement

1. The shipped goods have reached its final destination if the processing entity's expected last destination is equal to the last entity received. From this, the RFID details are updated according to the processing entity details.
2. Next phase can be continued if the policies of the PolicyChecker have been satisfied.
3. For a consumer to view transaction details, permission must be given by the processing entity of that particular role.

5.3 Multi-Transporters

A PoD system and framework for multi-transporters is proposed by Hasan and Salah [1]. They use existing systems like incentives and penalties to force carriers to be honest. They also include policies like refunds in case of delays. Contrast to the use of smart escrow contracts to ensure compliance; they make use of off-chain arbitrators where the arbitrator is given control over the money deposited as collateral relinquishing it after the settlement of a dispute.

For multi transporters in a smart city, the transporters are incentivized to act honestly by depositing collateral by each processing entity, ensuring trust among the chain of courier services (CSs) and transporters and providing a means of resolving disputes. The seller, buyer CS(s), arbitrator, certifying authority or smart contract attestation authority form a long chain for the system (SCAA).

The transporters sign an agreement between themselves and the buyer as well as the seller before transactions are initiated. Upon agreement set, the agreed-upon collateral is withdrawn from each entity (the buyer, the seller and the multi transporters).

Three types of contracts are created based on the need. Each child contract points to the next contract, creating a parent to the child contract system. Each child having the address of its parents and each parent having the address of their child references this. The initial contract that started the main chain is also referenced. This address is kept by all child and parent nodes. It is imperative to note that the supply chain should have at least two contracts. For instance, if one transporter is moving the physical asset, the only two smart contracts will be used. We therefore create three types of contracts known as PoD, Buyer to Transporter (BT) contract and CS. The PoD contract serves as the main smart contract. The BT serves as the smart contract at the end of the chain. For each supply chain if the transporters are less than one, then more than one chain of CSs is established.

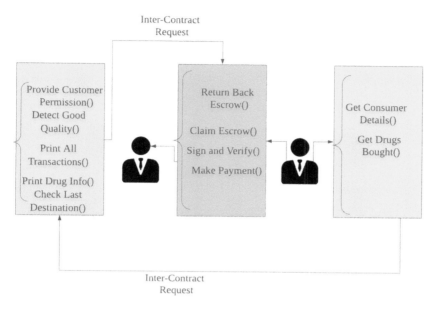

Inter-Contract
Request

Provide Customer
Permission()
Detect Good
Quality()
Print All
Transactions()
Print Drug Info()
Check Last
Destination()

Return Back
Escrow()

Claim Escrow()
Sign and Verify()
Make Payment()

Get Consumer
Details()

Get Drugs
Bought()

Inter-Contract
Request

FIGURE 5.5
Smart contract interaction.

5.3.1 Multi-Transporters Design

Three forms of contracts are designed to supply the item between the vendor
and the customer [9, 10]. The contracts as seen in Figure 5.5 are generated
based on the necessity and form a chain. Each contract leads to the next. So
every parent contract has the address of its child contract, and vice versa.
Also, every contract has the address of the first main contract [11]. The primary contract has an extra address: the last contract in the chain. It should
have two contracts. So, if just one transporter is required, two contracts will
be created. If more than one carrier is required, three contracts are made. It
also always starts with a PoD contract. So PoD is the core contract and BT is
the end of chain. If the number of transporters [12] is larger than one, then CS
contracts are made as needed as shown Figure 5.6.

Figure 5.7 demonstrates the contract chain with only one transporter. Each
contract in the chain is a PoD main contract and a BT end of chain contract.

5.3.2 The Seller

The seller deals with the PoD contract. Meanwhile, the CS contract and the
purchase contract are shared by the transporter and the buyer (BT). The PoD
contract specifies these contracts.

Notably, the required number of couriers is always one less than the
required number of transporters. The seller creates the package and hands

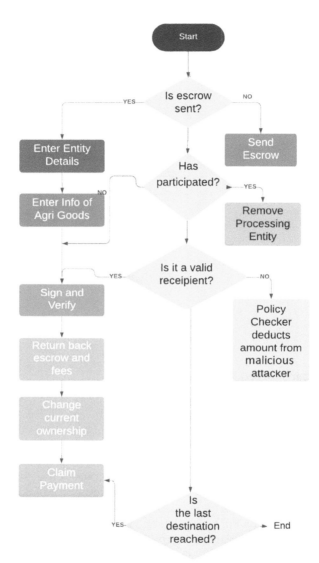

FIGURE 5.6
Flow diagram of agri-goods.

it over to the transporter physically. The transporter then creates the first CS contract. This CS [13] contract created by the first transporter has its terms and conditions agreed to by the second transporter. Upon verification with the key, the goods are handed over to the second transporter in a "hand-shake" manner. The second transporter notifies everyone that transporter one has arrived. If the verification process is complete and is successful, another CS contract is created by the second transporter. This process is repeated until

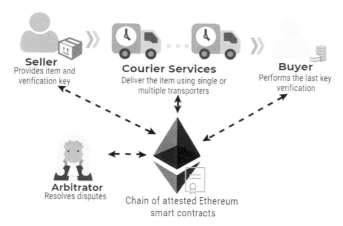

FIGURE 5.7
Multi-transporter transactions of goods over the blockchain network.

the buyer's address is reached. If the buyer's address is the destination, the BT and supply chain ends with the smart contract.

5.3.3 Signing and Terms and Conditions

Whenever a new contract is created on the blockchain, entities involved begin to sign the agreement involved. The contract specifies information about the trip to be pursued by the transporters. But no terms and conditions are signed for the last contract. Whenever the signing is completed, collateral is deducted from the signer's account as seen in Figure 5.8.

5.3.4 Cancellation and Funding

It is a challenge to cancel a delivery once it has started, but certain conditions are provided. Embedded within the contract is a Boolean cancellation state as well as the address of the caller of the cancel function. The seller can only

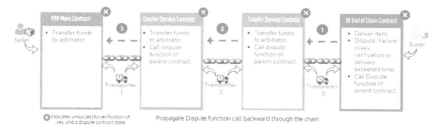

FIGURE 5.8
Contract-to-contract interaction of supply chain information of goods in a smart city.

FIGURE 5.9
Contract-to-contract interaction of supply chain information of goods in a smart city.

make the cancellation if the item is on its way to the next transporter. The buyer can also cancel if the item has not reached the next destination. Upon arrival, however, the cancellation will not be permitted as seen in Figure 5.9.

5.3.5 Expected Delivery Time

Since punctuality is essential for the efficiency of the delivery of goods on the supply chain. In the suggested paradigm, the buyer might decline an item if the carriers' delivery time exceeds the expected delivery time.

5.3.6 Payment Settlement

Upon verification, the last transporter delivers the goods to the buyer. In the BT contract, the keys and item are handed over to the buyer upon confirmation of the arrival of the transporter. The buyer then confirms the signature using his key, and the payment is instantiated.

5.4 Supply Chain Equilibrium on a Game-Theory-Incentivized Blockchain Network

The supply chain is made up of various players who all work together to ensure the safe delivery of products and services. These parties include organizations and individuals involved in the supply chain. This includes suppliers, retailers, and transporters who move goods throughout the supply chain. It is on these individuals to ensure the safe delivery of products and services. Sold goods and services may include manufacturing and manufactured goods and services [14]. However, these operations are costly, and how they are performed can directly impact each node's profit and the supply chain's overall profit. Participants in the supply chain compete to cut costs and maximize profit. For example, some retailers send large quantities of goods to cut shipping costs per item. Others to assure swift and reliable delivery of goods and services also provide users with a variety of shipping alternatives, giving them a competitive advantage [15].

However, this act can have a dire consequence on the profit of all individuals and the whole supply chain as a whole. Without the correct coordination mechanism, each node will play selfishly and in their selfish act, unbalance the effective distribution of profit for all parties on the chain. Since it is difficult to monitor a party's activity in the supply chain, it is very difficult to predict and monitor the actions of parties in the supply chain, and as such on the supply chain, individuals can play [16] to their fair advantage without the consideration for the rest of the members on the supply chain. Since all members do not trust each other and have no idea about the next move of the other participant on the supply chain, it is logical for an individual to avoid losses. He plays a strategy that will prevent him from making losses. This lack of trust and ignorance of the actions of members prevents coordination on the supply chain and prevents a Nash equilibrium from being achieved.

Traditionally, the Nash equilibrium has been established by utilizing GT models to estimate individual action payoff. It also forecasts whether an action will result in an optimal reward for all parties. The basic goal of GT is to define alternative outcomes and devise tactics to achieve them. These results are known as the Nash equilibrium. A Nash equilibrium is a reward where no member of the chain can improve their outcome by changing their strategy. After a while, the game enters a Nash equilibrium [17]. The model's main goal is to attain Nash equilibrium in time t. For a few parameters, the predicted payout of each action is predictable. A vast supply chain with complex participants and activities requires an automated technique of presenting a Nash equilibrium to the members. Since more than one Nash equilibrium is reached, the game cannot reach a Pareto optimal level. Automated systems must be able to manage complex procedures, for example. The blockchain, with its smart contract, also enables transparency and trust among members, as it connects via sensors and different IoT devices on the supply chain network.

Most supply chains, however, include a Stackelberg game, in which one player chooses a strategy first, then the other. Transparency among participants [18] ensures precise decisions that promote coordination among parties. With the transparency granted, we utilize Markov decisions to observe the game over time t. Our goal is to observe the optimal decision that will build a Nash equilibrium and secure profit for all parties involved in the supply chain. As a result, the blockchain provides access to verifiable and transparent actions. Although a lot of work has been done in blockchain and GT, we take a present a novel approach in its effectiveness on the supply chain and its ability to ensure fairness in a smart city. Utilizing GT and blockchain features to provide profitability to all members on the supply chain through a GT-incentivized network.

5.4.1 A Game-Theory-Based Supply Chain

Snyder and Shen [19] describe a supply-chain-based GT that involves strategic interaction between two or more players. This set of rules and outcomes is ideal for industry analysis and inter-firm interactions. Because GT assumes reason, it also assumes that each action will maximize a player's payoff. A dynamic [20] set of actions and payoffs are created by the numerous activities chosen in the game [21]. Using dynamic GT, participants continue to play after starting a round.

5.4.2 Iterated Prisoners Dilemma

Assuming the Prisoners' Dilemma is replayed in a two-stage game, we employ backward induction in the game like any dynamic game. Since no other method may be used, the game's outcome is attained at that level. Like any dynamic game, we solve it via induction. In the final stage of the game in Figure 1.2, because the best response is for Player 1 (P1) to play D irrespective of the Player 2 (P2's) strategy, (D, D) becomes the Nash equilibrium thus Player 1 has a payoff of 1. The sub-game in itself is not the first stage. The sub-games starting from the first game can be considered to be the game. The payoff for the entire game may be determined by adding the payoffs for the Nash equilibrium at the last stage to the payoff at the first stage. The pure-strategy set for each player in the entire game is S = {CC, DC, CD, DD} but because we are only interested in a sub-game perfect Nash equilibrium, we only need to consider a subset of the payoff table.

5.4.3 Repeated Games

In the Prisoners' Dilemma, Haskell, and Jain [22] notice a pattern of uncooperative behavior to achieve a payout. The game's payout increases dramatically if played cooperatively. Uncooperative behavior is expected, even though cooperative behavior leads to better rewards for all players [23]. Firms usually make multiple decisions. The players' payout is based on past decisions since interactions are based on earlier choices that put them in a different state. So the next action will change their condition. A stage game is a game where only one state and one decision are played. After the game, the players are faced with a new set of decisions to enter a new state. The game is repeated. The most crucial approach is to always establish Nash equilibrium for the stage game. Players make decisions based on other players' decisions and threaten future actions if the game's route does not match their desired plan and decision. The payout is greater when the game is regarded as a whole, as shown by Haskell and Jain in [24].

Theorem 1
Every game with a finite strategic form has at least one Nash equilibrium.

TABLE 5.1

A Payoff Table Showing a Subgame
Perfect Nash Equilibrium

		P2	
		CD	DD
P1	CD	4,4	1,6
	DD	6,1	2,2

Supposing the Prisoners' Dilemma is repeated in a two-stage game, we use backward induction in the game just like any dynamic game. Because there will be no more strategies to play at the last stage of the game, the game's outcome will be determined at that point. Backward induction is used to solve this, just like any other dynamic game. Because the optimum answer is for Player 1 (P1) to play D regardless of Player 2 (P2's) approach, (D, D) becomes the Nash equilibrium at the final stage of the game, as shown in Table 5.1, and Player 1 obtains a reward of 1. The sub-game is not the initial stage in and of itself. Starting with the first game, the sub-games can be regarded the game.

However, because the strategies are designed for the game's ultimate stage, the payout may be determined by adding the payoffs for the Nash equilibrium at the final stage to the payoff at the first stage to get the overall reward. The pure-strategy set for every player in the entire game is S = CC, DC, CD, DD; however, we only need to evaluate a subset of the payoff table because we are only interested in a sub-game perfect Nash equilibrium.

5.4.4 Markov Decision Process

The Markov decision process (MDP) is a method for dynamically controlling stochastic systems. This is the fundamental structure. We concentrate on discrete-time models, in which we see the system at moments $t = 1, 2,..., n$, where n is the horizon, which might be finite or infinite. By selecting particular criteria known as actions at each time unit, a player can influence the game's cost and evolution. The behavior varies depending on the system's current state and the system's control action. These are what we refer to as the system's "state." By selecting particular parameters, called actions, at each time unit, a controller can impact both the costs and the evolution of the system. We assume, as is commonly the case in Control Theory, that the system's behavior at any given time is governed by the system's "state" as well as the control action. The system transitions from one state to the next in a random order. The likelihood of shifting to the next state at time t is determined by the current state of the system and the control action. Markov chains have some interesting qualities. For example, an important Markov condition is that the previous state and the new state are all independent at any point between the state s and the action t, as well as between the following $s + 1$ and $t + 1$.

We define a tuple $\{X, A, P, x, d\}$, which is a state space containing a finite number of states. Let state denote x, y, z. A is a finite set of actions. At A we define $A(x) \subset A$ as those actions available at x. Let $K = (x, a): x \in X, a \in A(x)$ be the set of state-action pairs with a being a generic notation for an action. P is the transitional property, thus the probability of moving from state x to y if action a is chosen is denoted as $P(y \mid x, a)$.

5.4.5 Stackelberg Model

In this chapter, we apply a Stackelberg game technique to create a leader-follower game with predefined strategies. The leader initiates the first motion, followed by the follower. The follower decides on a strategy after seeing the leader's strategy. The two players' main goal is to maximize their profit, thus they should act rationally. For instance, two competing firms ($i = 1, 2$) selling divisible products must decide how much of it they would want to produce. Taking both players as P_1 and P_2, we denote q_i as the amount of quantity of each firm. We assume that the market price of the product is given by $P(Q) = P_0 \left(1 - 1\dfrac{Q}{Q_0}\right)$, where $Q = q_1 + q_2$. c denotes the cost of unit of production for each firm. The Stackelberg game is played successively. For example, Firm 1 (the leader) decides on a quantity to create, and Firm 2 (the follower) decides on an amount q to produce. Assuming each firm wants to maximize its profit, that is $P_0 > c$, in order to have a sub game perfect Nash equilibrium, Firm 2 makes a best response $\hat{q}_2(q_1)$ for every possible choice of production quantity set by Firm 1. The game is solved by backward induction to determine the maximum payoff $\pi_1(p_1, p_2), \pi_2(p_1, p_2)$ both players gain to achieve sub-game perfect Nash equilibrium. Therefore Player 1 chooses his strategy in order to achieve a profit of $\pi_1(p_1, \sigma_2)$ and Player $2\pi_2(p_2, \sigma_2)$, does likewise in a leader-follower strategic pattern. Both players play until a Stackelberg equilibrium and thus at the same time solving the optimization problems.

5.5 Game-Theory-Incentivized Blockchain-Based Network Overview

In this section, from Figures 5.10 and 5.11, we demonstrate our blockchain-based supply chain system that uses GT and provides coordination for all parties. We apply the Iterated Prisoner's Dilemma and the outcome in Markov decisions to the blockchain to achieve an appropriate outcome over time t. In other circumstances, the parties can agree to any supply chain contract, however for our work, we employ the wholesale price supply chain contract. We also study the interaction [18] of retailers R and suppliers S.

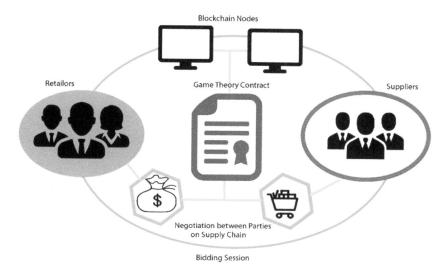

FIGURE 5.10
Bidding system showing an interaction between retailer and supplier on a game-theory-incentivized blockchain network.

We created a GT smart contract to assure blockchain party coordination [25]. A retailer R_i contacts a group of suppliers S on the supply chain to purchase a certain good G. A bidding process is as seen in Figure 5.12 is established to ensure that all parties are treated fairly and that the supply chain as a whole is profitable. Each supplier, therefore, takes an optional action a_1, a_2 whether to participate in the bid or not. If they do agree, the bidding process is initiated. For any of the actions a_i from a set of actions A, taken the expected

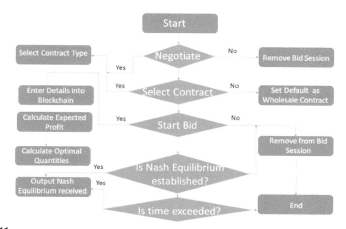

FIGURE 5.11
Flow diagram showing interaction between retailer and supplier to establish Nash equilibrium on the blockchain.

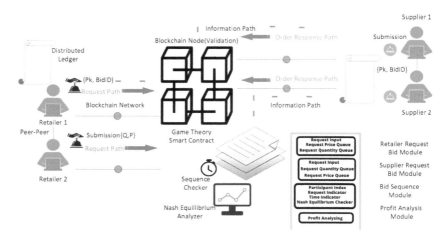

FIGURE 5.12
Diagram showing interaction among blockchain-based IoT devices on the supply chain network.

payoff is denoted $\pi(a)$. The maximum payoff is thus $a^* \in \arg max\ \pi(a), a \in A$, where $\pi(a^*) \geq \pi(a), \forall a \in A$.

At each point on the blockchain for an action set A_i the number of actions n is given by $A_1 \times \cdots \times A_{n-1} \times A_n$; it shows the set of pure strategies for all the nodes on the blockchain for each type of role played by individuals on the supply chain.

All participating retailers and suppliers enroll in a bid with a unique id, bid_{ID}. Their blockchain addresses that involve a public key pk, names, and roles are entered and stored unto the blockchain. Thus, all interactions involved will be associated with the bid set up between the retailer R_i and the suppliers S_i. The retailer's expected profit function can be given as

$$\pi_r(Q) = (r - v + p_r)S(Q) - (c_r - v)Q - p_r\mu - T \tag{5.1}$$

Supplier S_i also decides to also propose his bid based on the decision made by retailer R_i. His per unit cost in supplying p_s is calculated as

$$\pi_s(Q) = p_sS(Q) - c_s(Q) - p_s\mu + T \tag{5.2}$$

The smart contract considers all the values involved and thus produces the optimal payoff that will be essential for both parties. Taking into consideration a chain involving n participant at time t, this can be calculated as

$$\Pi(Q) = (r - v + p)S(Q) - (c - v)(Q) - p\mu \tag{5.3}$$

This is automatically added to the blockchain. Other pertinent data is saved in a P2P network like IPFS.

The RetailOptimalOrderQuantity() function calculates the least optimum quantity from the list of quantities Q1,..., Qn proposed by. The SupplyOptimalOrderQuantity() function calculates Q_s^* for each supplier S. The SupplyOptimalOrderQuantity() function calculates the least optimal quantity among the suppliers in S. On the blockchain, the optimal Qn for R and S members is calculated as follows:

$$Q_r^* = \frac{w + c_r - v}{r - v + p_r} \tag{5.4}$$

$$Q_s^* = \frac{w - c_s}{p_s} \tag{5.5}$$

The ideal amount for the entire supply chain is then computed as Q^0:

$$Q^0 = \frac{c - v}{r - v + p} \tag{5.6}$$

The game is played several times from time t_0 to time t_n by all the blockchain nodes associated with a particular bid_{ID}, until the particular Q^0 and $\pi(Q)$ (the particular quantity that optimizes profit for the whole supply chain) is achieved.

The blockchain is said to be coordinated if the $Q^0 = Q_s^* = Q_r^*$ among all selected quantities provided by all the parties [1] in that particular bid. In other words, the optimal quantity discussed serves as the right number of commodities to be given, benefiting all parties concerned.

Theorem 2

$Q_r^* = Q_s^* = Q^0$ if and only if $w = cs - \dfrac{c - v}{r - v + p} ps.$

$$Q_s^* = \frac{w + c_r - v}{r - v + p_r}$$

$$= \frac{\left(c_s - \dfrac{c - v}{r - v + p} ps\right) + c_r - v}{r - v + p_r}$$

$$= \frac{(c - v)\left(1 - \dfrac{p_s}{r - v + p}\right)}{r - v + p_r}$$

$$= \frac{c - v}{r - v + p}$$

$$= Q^0$$

Since Q is strictly decreasing and continuous, this implies $Q_r^* = Q^0$. Similarly,

$$Q_s^* = \frac{c_s - w}{p_s}$$

$$= \frac{c_s - \left(cs - \dfrac{c - v}{r - v + p} p_s\right)}{p_s}$$

$$= \frac{c - v}{r - v + p}$$

$$= Q^0$$

This theorem shows that the smart contract can calculate a value w that resolves to $Q_r^* = Q_S^* = Q^0$.

5.5.1 Game Theory Analysis

Suppose a node [26] R as seen in Figure 5.13 acting as a retailer on the blockchain with a set of actions $A = a_1, a_2, \ldots, a_n$ chooses a random action as the optimal response to each action supplied by S, he will choose an action that will increase or maximize his payoff. For example, he can raise the selling price of the goods to be bought and set a high value of p, or he decides to buy the goods in bulk. It is quite difficult to predict the next behavior β that will be taken by R for the rest of the game since human decisions are based on consideration and abounded rational behavior that comes together with randomness and emotions. We will therefore account

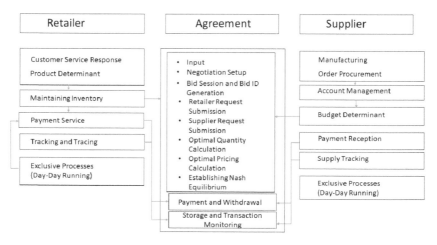

FIGURE 5.13
Proposed contracting mechanism for multiple entities on the supply chain network.

for the probability that R will choose a particular action β to obtain a payoff of $\pi(\beta \mid x)$ as

$$\pi(\beta) = \sum_{\alpha \in A} p(a)\pi(a) \tag{5.7}$$

Due to the randomness [27] and uncertainty in R choice of actions, for each action in a continuous strategy game we produce an expected payoff of $\pi(\beta) = \sum_{x \in X} P(X = x)\pi(\beta \mid x)$.

Thus, as the game continues, R will take an expected behavior of $\beta^* = \ _{\beta \in B}\ \pi(\beta)$, providing an optimal payoff of $\pi(\beta^*)$ for R, which optimizes the whole supply chain. Players may need to look at the game randomly to make the right selection. Because multiple actions $a_1, a_2, \ldots, a_n$ might lead to optimal outcomes, the game's randomization can reveal the best outcome. Using the ExpectedRetailerCost(), ExpectedSupplyCost() and TotalExpectedProfit() functions, we may arrange the payoffs from lowest to highest.

Using the blockchain, we can discover the optimal supplier whose proposal and quantity will yield the desired profit [28] on the supply chain. We prove here that there exists an action a_i that can give a payoff of more than $\pi(\beta^*)$ contradicting the original idea that $\pi(\beta^*)$ always provides the maximum payoff.

Theorem 3

Suppose there exists a set of optimal actions A^*, then $\pi(a) = \pi(\beta^*), \forall a \in A^*$.

$$\pi(\beta^*) = \sum_{a \in A^*} p^*(a)\pi(a)$$

$$= \sum_{a \neq a'} p^*(a)\pi(a) + p^*(a')\pi(a')$$

$$< \sum_{a \neq a'} p^*(a)\pi(a') + p^*(a')\pi(a')$$

$$= \pi(a')$$

5.5.2 Transparency

The blockchain provides transparency. A distributed ledger maintains consensus across all nodes. Transactions are hashed and kept in an unchangeable ledger. This influences the decisions made, since the game depends on each agent's knowledge of the other and how well their actions and judgments are understood. The blockchain will allow high levels of interaction among all stakeholders, high levels of payoffs, and effective decision-making.

5.5.3 State Change

A timestamp is kept, which indicates at what time each state x on the block-chain was changed. Every strategy considered at any time is denoted as $a_t = a(x_t, t)$. The current state x of the blockchain induces an action $x_t \to x_{t+1}$. Each time the actions performed by all the players up to that time on the blockchain are denoted as $h_t = (x_0, a_0, x_1, a_1, \ldots, x_t)$ and is stored and hashed unto the blockchain.

In an infinite game, the probability of a certain state X on the blockchain based on player actions is

$$\sum_{x_{l+1} \in X} p(x_{t+1} \mid h_t, a_t) = 1 \tag{5.8}$$

Using a randomized technique, his estimated return is

$$\pi(x \mid \sigma) = E\left[\sum_{t=1}^{T} r_t(X_t, A_t) + r_T(X_T)\right]$$

$$= \sum_{t=1}^{T} E[r_t(X_t, A_t)] + E[r_T(X_T)] \tag{5.9}$$

The key goal is to make a tactical decision over time $t+1$ and find the best answer. Taking the dynamic programming equation as

$$\pi_t^*(x) = \max_{a \in A} \pi_t(a \mid x)$$

$$= \max_{a \in A}\left[r_t(x, a) + \sum_{x' \in X} p(x' \mid x, a)\pi_{t+1}^*(x')\right] \tag{5.10}$$

Extending our prediction toward time independence, assuming for a time $t = 0$ to an infinite $\lim it(\infty)$, all players take a strategy σ, for instance, the optimal strategy for a set of continuous strategy will result in an optimal payoff of

$$\pi^*(x) \equiv \pi(x \mid \sigma^*)$$

$$\leq \max_{a \in A(x)}\left[r(x, a) + \delta\sum_{x' \in X} p(x' \mid x, a)\pi^*(x')\right] \tag{5.11}$$

Regardless of the state, after each decision is made, each node will be faced with the same infinite horizontal decision problem as previously. Therefore, it shows that for every quantity of goods that will be produced $X + 1$, the strategy that will be played with a probability of P achieves a payoff $\pi(x')$.

The optimal payoff obtained meets the dynamic programming equation, proving that, regardless of the infinite time on the blockchain, each node can reach a payoff of and a maximum payoff at any time, each node can achieve an payoff of $\pi(x)$ and a maximum payoff π^*.

5.5.4 Infinite Games

Suppose that the player R_i receives a payoff of $r_i(t)$ in each stage t, then their total payoff of at the end of time ∞, that is payoff at the end of each game will be $\sum_{t=0}^{\infty} r_i(t)$.

However, it is a bit simplistic to believe that simply glancing at the end stage will reveal each player's eventual payment, because stationary methods may differ for each player at each level.

5.5.5 Establishing Nash Equilibrium

Assume that the discount factor symbol in the inequality is bigger than half. $\delta \geq \frac{1}{2}$, a Nash equilibrium [29] can be established since the discount factor is high enough. However, if both players decide to play differently, then we know that σ_R, being played by one player, will get a different payoff from σ_i.

5.6 Conclusions

In this, a method to ensure PoD for goods is looked at using blockchain features; we develop a mechanism to monitor these goods as they are moved from one party to another. We look at ways of improving the verification of goods. They are handed over by one party to the other and easily provide receipts and payment after delivering goods from one end of the chain to the other. A blockchain-based architecture is also designed to show the idea. In the architecture, the permission-based feature is provided, which enables the processing entity to either decline or provide access to consumer using re-encryption keys. Each transaction is provided to prevent the other entities such as competitors from gaining access to the system.

We also looked at using blockchain to make the supply chain profitable for all participants. We used Iterated Prisoners Dilemma [21] between suppliers and retailers in a game. For the blockchain-enabled supply chain, we studied the game to determine when Nash equilibrium is reached. A mathematical framework for creating Nash equilibrium utilizing the blockchain was developed [20, 24]. In future works, we will look at ways to maintain profitability on the supply chain for all stakeholders on the blockchain using evolutionary GT.

References

[1] Hasan, H. R., & Salah, K. (2018). Proof of delivery of digital assets using blockchain and smart contracts. IEEE Access, 6, 65439–65448.

[2] Toyoda, K., Mathiopoulos, P. T., Sasase, I., & Ohtsuki, T. (2017). A novel blockchain-based product ownership management system (POMS) for anti-counterfeits in the post supply chain. IEEE Access, 5, 17465–17477.

[3] Caro, P., Ali, M. S., Vecchio M., & Giaffreda., R. (2018). "Blockchain-Based Traceability in Agri-Food Supply Chain Management: A Practical Implementation," in 2018 IoT Vertical and Topical Summit on Agriculture – Tuscany (IOT Tuscany), Tuscany, pp. 1–4.

[4] Bo, Y., & Danyu, L. (2009). "Application of RFID in Cold Chain Temperature Monitoring System," in 2009 Second ISECS International Colloquium on Computing, Communication, Control, and Management, CCCM, vol. 2, pp. 258–261.

[5] Zhang, R. (2013). "Applying RFID and GPS Tracker for Signal Processing in a Cargo Security System," in: 2013 IEEE International Conference on Signal Processing, Communication and Computing (ICSPCC 2013), KunMing, pp. 1–5.

[6] Albuquerque, B. & Callado, M. (2015). Understanding Bitcoins: Facts and Questions. Revista Brasileira de Economia. 69. https://doi.org/10.5935/0034-7140.20150001

[7] Solidity Documentation 0.5.3. (2019). https://media.readthedocs.org/pdf/solidity/develop/solidity.pdf

[8] Wood, G., & Buterin, V. (2014). Ethereum: A secure decentralised generalised transaction ledger. Ethereum Project Yellow Paper. https://doi.org/10.1017/CBO9781107415324.004

[9] Kosba, A., Miller, A., Shi, E., Wen, Z., & Papamanthou, C. (2016). "Hawk: The Blockchain Model of Cryptography and Privacy-Preserving Smart Contracts," in: 2016 IEEE Symposium on Security and Privacy (SP), San Jose, CA, pp. 839–858.

[10] Koblitz, N., Menezes, A., & Vanstone, S. (2000). The state of elliptic curve cryptography. Designs, Codes and Cryptography, 19, 2–3, 173–193.

[11] Singla A., & Bertino, E. (2018). "Blockchain-Based PKI Solutions for IoT," in: 2018 IEEE 4th International Conference on Collaboration and Internet Computing (CIC), Philadelphia, PA, pp. 9–15.

[12] Yu, S., Lv, K., Shao, Z., Guo, Y., Zou, J., & Zhang, B. (2018). "A High Performance Blockchain Platform for Intelligent Devices," in: 2018 1st IEEE International Conference on Hot Information-Centric Networking (HotICN), Shenzhen, pp. 260–261.

[13] Behera, S., & Maity, C. (2008). "Active RFID tag in Real Time Location System," in: 2008 5th International Multi-Conference on Systems, Signals and Devices, Amman, pp. 1–7.

[14] Xie, W., et al. (2018). "ETTF: A Trusted Trading Framework Using Blockchain in E-commerce," in: 2018 IEEE 22nd International Conference on Computer Supported Cooperative Work in Design (CSCWD), Nanjing, pp. 612–617.

[15] Xiong, Z., Feng, S., Wang, W., Niyato, D., Wang, P., & Han, Z. (2019, June). Cloud/fog computing resource management and pricing for blockchain networks. IEEE Internet of Things Journal, 6, 3, 4585–4600. https://doi.org/10.1109/JIOT.2018.2871706

[16] Dey, S. (2018, September). "Securing Majority-Attack in Blockchain using Machine Learning and Algorithmic Game Theory: A Proof of Work," in: 2018 10th Computer Science and Electronic Engineering (CEEC), IEEE, pp. 7–10.

[17] Pokrovskaia, N. N. (2017). "Tax, Financial and Social Regulatory Mechanisms within the Knowledge-Driven Economy. Blockchain Algorithms and Fog Computing for the Efficient Regulation," in: 2017 XX IEEE International Conference on Soft Computing and Measurements (SCM), St. Petersburg, pp. 709–712.

[18] Kim, S. (2019). Two-phase cooperative bargaining game approach for shard-based blockchain consensus scheme. IEEE Access, 7, 127772–127780.

[19] Snyder, L. V., & Shen, Z. J. M. (2011). Fundamentals of supply chain theory (p. 367). Hoboken, NJ: Wiley.

[20] Pardalos, P. M., Migdalas, A., & Pitsoulis, L. (Eds.). (2008). Pareto optimality, game theory and equilibria (Vol. 17). Germany: Springer Science & Business Media.

[21] Bigi, G., Bracciali, A., Meacci, G. & Tuosto, E. (2015). Validation of decentralised smart contracts through game theory and formal methods. In: Programming languages with applications to biology and security (pp. 142–161). Springer, Cham.

[22] Thun, J. H. (2005). The potential of cooperative game theory for supply chain management. In: Research methodologies in supply chain management (pp. 477–491). Physica-Verlag HD.

[23] Gao, H., Ma, Z., Luo, S., & Wang, Z. (2019). BFR-MPC: A blockchain-based fair and robust multi-party computation scheme. IEEE Access, 7, 110439–110450. https://doi.org/10.1109/ACCESS.2019.2934147

[24] Haskell, W. B., & Jain, R. (2012, December). "Dominance-Constrained Markov Decision Processes," in: 2012 IEEE 51st IEEE Conference on Decision and Control (CDC), IEEE, pp. 5991–5996.

[25] Eyal, I. (2015, July). "The Miner's Dilemma," in: Proceedings – IEEE Symposium on Security and Privacy, pp. 89–103. https://doi.org/10.1109/SP.2015.13

[26] Liu, Z., Luong, N. C., Wang, W., Niyato, D., Wang, P., Liang, Y. C., & Kim, D. I. (2019). A survey on blockchain: a game theoretical perspective. IEEE Access, 7, 47615–47643.

[27] Narayanan, A. (2018). "Blockchains: Past, Present, and Future," in: Proceedings of the 37th ACM SIGMOD-SIGACT-SIGAI Symposium on Principles of Database Systems, p. 193. https://doi.org/10.1145/3196959.3197545

[28] Manshaei, M. H., Jadliwala, M., Maiti, A., & Fooladgar, M. (2018). A game-theoretic analysis of shard-based permissionless blockchains. IEEE Access, 6, 78100–78112.

[29] Song, L. H., Li, T., & Wang, Y. L. (2019). Applications of game theory in blockchain. Journal of Cryptologic Research, 6(1), 100–111. https://doi.org/10.13868/j.cnki.jcr.000287

6

Smart Transportation

6.1 Introduction

Smart transport is essential to the smart city's operation. Complex road networks and sophisticated transportation systems are linked to traffic safety, traffic control, and smart parking. The main frameworks for a smart transportation system consist of route prediction and sensing distribution and execution for intelligent transport networks. Various innovations have emerged to facilitate smart transport. For example, technologies that assure traffic safety and monitor animal traffic within a designated safety zone have been developed. Sensor devices can also warn drivers of potential hazards such as obstructions, animals, poor road conditions, or vehicles traveling in the wrong way. These gadgets communicate to alert drivers of impending occurrences, assuring their safety.

A look at smart cities transportation aims to minimize transportation problems associated with smart cities and look at existing technologies and innovations that help resolve major issues in smart cities. Congested roads and vehicular motor accidents remain a major issue in smart cities. They are essential for current designs and solutions to be tailored toward ensuring smart cities are effectively network, and safe transportation provided across all points. With the growth of smart cities population, individuals' daily movement in a smart cities becomes more complicated. As such, an effective way of ensuring mobility in smart cities is needed. They are solving these problems through digital services, sensors, smart appliances, and infrastructure such as road networks as well-connected devices help provide suitable mobility for all entities in the smart cities.

6.2 Framework for Transportation

In [1], the authors propose a paradigm for addressing big data privacy and security issues in smart transportation. Individuals, businesses, governments, and universities collect and control their data. Participants encrypt their data onto the blockchain and agree to the regulations set by the block-chain owner. They provide a rigorous approach to ensure ownership, transparency, and auditability of data using the blockchain. Since it is easier to analyze and process data due to the availability of more advanced technological systems, it is now easier to understand human mobility for designing smarter, demand-driven, reliable, and secure transportation systems [2].

The blockchain network's nodes continuously create mobility data under this architecture. Several companies are working on providing information that will be essential to governments and researchers. These include telecommunication networks that provide data needed for transportation modeling and logs on individuals' mobile devices on a road network to monitor traffic. These companies' data sources can help improve transportation networks. The data on tolls, smart cards, and traffic detectors can be used to shape smart city mobility [3]. Smart card fares can be forecasted using data from traffic detectors and these devices. Other information that is collected by governments and universities can be used to detail car emissions or monitor driver behavior.

Of course, the largest contributors of data are the individuals and smartphone owners. Researchers can use this information provided to provide cutting-edge transport solutions. It can help the government also to plan and build better transportation systems. It can also assist companies in delivering customized products and services.

This framework creates smart contracts as seen in Figure 6.1. Their role is to accept connections [4], allow access to specific information, and revoke connections. In the previous diagram, smart contracts are utilized to share personal data between two nodes. The nodes have full access to the area of the ledger that includes their data. In this framework:

1. Connection to a smart contract is requested by a receiving node.
2. The smart contract verifies that the receiving node has the information and accepts it.
3. The nodes then begin sharing information with one another on a peer-to-peer basis.
4. Only if the connection is found on the smart contract can the shared data be accessed.

With this framework, they resolved issues associated with the following:

1. *User privacy*: By making sure users hold their data privately and created a system where individuals feel safe when sharing their information.
2. *Data ownership*: They provide a system where individuals can leave the network and revoke access to their data by any other entity on the blockchain.

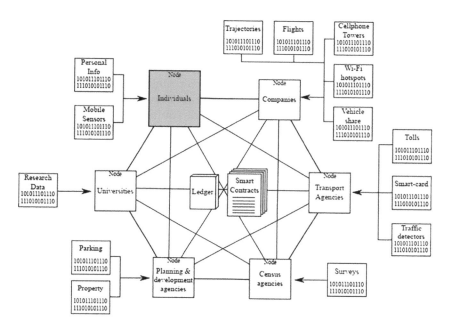

FIGURE 6.1
Smart contract and transaction.

3. *Data transparency and auditability*: Individuals can access areas of the ledger where their data is stored, ensuring openness in the transfer of users' data. Due to blockchain's transparency, they can easily audit the data in this way.

4. *Fine-grained access control*: It might be difficult for businesses to withdraw access to their data at times. Companies and individuals can manage access to certain pieces of their information thanks to the blockchain's availability and inbuilt smart contracts.

5. *Data interception*: Attacking a single node will not be worth the work necessary to decode the data because all information exchanges take place across a unique and secure peer-to-peer link.

6. *Data leaks*: Massive hacks will require a huge amount of power due to the blockchain's decentralized nature to affect a meaningful number of individuals.

7. *Unsolicited share of information*: Because each node has access to the area of the ledger where their data is stored, they can simply verify the information they require.

8. *Unsolicited request of information*: The smart contract allows nodes to choose the information they want to exchange with other nodes based on their preferences.

6.3 Mobility as a Service

Others [5] propose a smart transportation system with Mobility-as-a-Service (MaaS) for edge computing. Currently, MaaS systems act as an intermediary layer that regulates and controls connections between transportation providers and passengers, with the objective of connecting this existing layer. We observe this in the form of blockchain-based MaaS, their solution, which combines transportation services, transportation information, and payment services and helps to promote confidence and transparency among all stakeholders. The blockchain eliminates the need for distinct MaaS agents. Decentralized computer power is dispersed to diverse transportation providers at the network's edge. A blockchain-based system for smart city mobility, focusing on MaaS, offers efficiency and reduces CO_2 emissions.

6.4 General Structure of MaaS

Participants in MaaS include transportation providers who operate as validators, and travelers who get route information from providers via a smart contract (based on their preferred).

The traveler will receive information that displays multiple means through which they might send a selected probable set of transit modes in the smart city, as shown in Figure 6.2. A smart contract is formed before responding to the network. Providers in blockchain-based MaaS try to verify and confirm the smart contract. To authenticate the smart contract from the traveler, the providers must examine the transit modes and needs. Then the blockchain gets it. As a result, transportation providers must share their routes to increase the likelihood of route sharing. Once shared, the routes are confirmed on the blockchain. They must, however, communicate information about their paths to increase their chances. A voting system is built to avoid uncovering bad suppliers' routes. Before sending this agreed-upon smart contract to their neighbors, each transportation provider must validate submitted smart contract based on data routes. Due to the smart city's voting system, transportation providers must share their knowledge of routes to gain other providers' consent. One of the primary questions in this blockchain-based system is how to close the system, that is, how to cease collecting certified smart contracts. Lots of proposals have been presented, whether to tackle a specific problem or to wait. After consensus, a set number of confirmed smart contracts must be obtained.

The classic framework has a server-side and client-side architecture. Transport providers as seen in Figure 6.3 can create candidates in the network. Due to the asynchronous nature of the blockchain network, each transportation provider can receive and verify different smart contracts. After

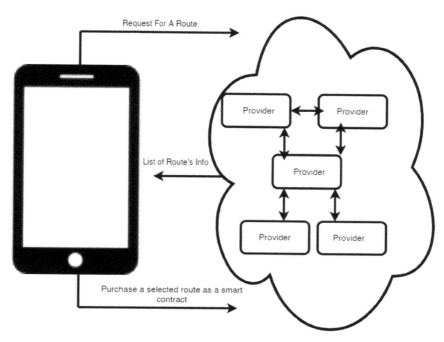

FIGURE 6.2
An overview of blockchain-based MaaS.

closing the current block, each transportation provider must broadcast the block to its neighbors to obtain consensus. As long as the transportation provider obtains half the threshold, the block is added. The MaaS also employs Bitcoin's incentive structure to secure voting unanimity among parties. After paying for a route, travelers must pay an additional cost to compensate providers for voting. Travelers must pay an additional fee after paying for a route to thank providers for participating in the route's voting process. There are two communication endpoints between client and server.

6.4.1 Ensuring Confidentiality

One of the things to address is user privacy. Since smart contracts must be automatically enforced, mostly the terms of the contracts are written broadly and openly. It is vital to ensure the safety of personal information while on the network.

6.4.2 Cryptographic Schemes

To preserve network privacy, it is critical that the contract obligations are met without compromising confidential information. Implementation security

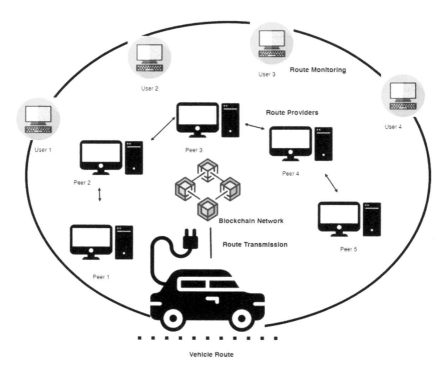

FIGURE 6.3
The specific communication between a traveler and a transportation provider.

can be ensured by using zero-knowledge argument schemes. Because the blockchain is a private network, users' privacy is an important factor that the system must consider.

6.4.3 Vulnerabilities of Smart Contracts

Another issue to note is the use of smart contracts. A case study of the blockchain's vulnerabilities with smart contracts as a case study shows a lot of vulnerabilities coming from the programming language and the blockchain itself. Most programmers cannot recognize the security problems or mistake, and as such, the smart contracts can be utterly flawed with a lot of vulnerabilities. These issues can be mitigated as long as the programmers understand basic security principles and have tools that enhance low-level static analyses of codes written.

6.4.4 Vital Role

Blockchain-based MaaS is an interesting concept since it provides trust between providers and ensures flexibility and transparency in a peer-to-peer network. Centralization is always an important aspect to consider when one thinks of a smart city. However, more work needs to be done to create better journey planners and protect smart contract's security.

6.4.5 Impact

MaaS is one of the best potential alternatives for smart transportation. The blockchain will enable transport operators to connect their services to complete journeys. Extra commercial agreements between service providers will be required. This facilitates governance, data sharing, route planning, and timing.

6.5 Incentive for Intelligent Transport Systems

Vehicular announcement networks are a crucial utility in smart vehicle communications. But there are two issues with a vehicle announcement network. It is challenging to send reliable messages without revealing user identities. Second, users aren't motivated to forward messages. Creditcoin [6] provides an effective vehicle announcement aggregation protocol. Without full trust, unknown signers or users can generate signatures and deliver announcements anonymously. But with blockchain, users might be rewarded for sharing traffic data. Because the blockchain is tamper-proof, transaction data is maintained immutable. It ensures conditional privacy by tracing harmful users' identities through anonymous announcements with linked transactions, as shown in [7].

It is one of the most potential vehicular communication applications in VANETs (Vehicular ad hoc Networks) and it makes driving safer. It also saves money by decreasing traffic bottlenecks and accidents. Blockchain-based networks are efficient in recording credit data and are decentralized, which is vital in VANET. Building an efficient vehicular announcement network has two key challenges due to data privacy considerations. First, all messages [8] in VANETs must be sent anonymously because they often contain sensitive information like vehicle numbers, driving preferences, and customer names. Notably, forwarding messages does not guarantee announcements [9]. Second, users find it difficult to forward messages via VANET if their privacy is at danger. Users get nothing from forwarding announcements. This decreases their motivation to respond.

Creditcoin is a blockchain-based VANET network incentive. Using protocols like Echo-Announcement for practical use in forwarding announcements. Users maintain reputation while spending Bitcoin as rewards on the blockchain. In the event of an unforeseen event, the blockchain provides conditional privacy. Figure 6.4 displays Creditcoin in general.

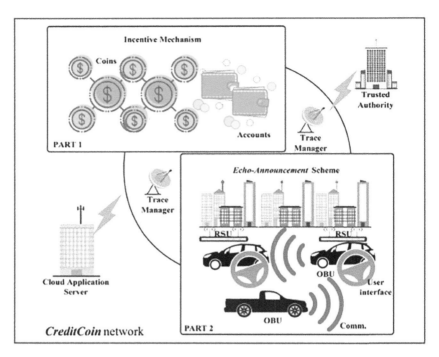

FIGURE 6.4
Overview of Creditcoin.

The reward method, Creditcoin, is simple. The program uses the announcement protocol Echo-mechanism to encourage users to forward true announcements. One of the proposed solutions is a Cloud application server with a Trusted Authority. Each user is granted a credit account that is linked to their reputation. Users choose between Replier (R) and Initiator (I).

A user acts as a R in the network. New users start with a few coins. At this moment, he can only respond to inquiries as R in the network. The user must stay active to post a lucrative quest for traffic information. If he or she has enough coins, they can shift roles and forward the requests later. He or she may do so for two reasons. One is to earn extra coins from a rewarding mission. The other is if he volunteers to inform others. Because submitting requests costs Bitcoin, this helps prevent useless requests. Once a user has reached the mission posting level, they can create a mission to gather traffic data.

Users are urged to reward additional coins. This keeps the network alive. In addition, the suggested Creditcoin system improves the availability and non-repudiation of vehicle announcements. A malicious attacker can't quickly change the voting system due to the blockchain. This mechanism stimulates coin flow. The Trace Manager's operation in this system also reveals the attackers' identities and addresses.

6.5.1 Proposition of Ideas

Creditcoin builds an incentive announcement network with a tokenized coin balance. The prepositional rules are as follows:

- *Proposition 1 (reply a QRP)*: When T broadcasts a request QRP, R receives numerous coins if R replies. Additionally, to prevent abuse of replies, a user's daily reply frequency is limited.
- *Proposition 2 (post a rewarding task)*: H will get information about a specific area if there is a relay of information in that area. H, therefore, creates a rewarding mission with attractive rewards.
- *Preposition 3 (finish a rewarding mission)*: For a user who becomes I and forwards the Anchor-Geography Based Routing Protocols (AGP) successfully, there is a degree of eagerness to receive a reward.
- *Preposition 4*: To send an announcement, I must first send a QRP request to other users. I must send coins before sending a QRP. If H hears the news, I get reward according to Proposition 3. According to Section 5, I should send RQPs to other users to forward an announcement. Spend coins before sending RQPs. The payoff usually outweighs the cost of sending RQPs. If I ask for something that isn't honest, few individuals will react. This reduces the I coins. Malicious users can't keep spamming the system. So the system and users are honest.
- *Proposition 5*: The unspent coins are halved in a certain period of time. This prevents coins from being accumulated and harnessed for attacks.

Note: *QRP, RQP are names of request sent on the network.*

6.6 Blockchain in Vehicular Systems

The existing vehicle Global Positioning System (GPS) lacks the accuracy required for autonomous driving. The existing vehicular GPS is far from accurate enough to support autonomous driving. Range and cooperation have been exploited to some degree to alleviate positional accuracy concerns, but it is still a challenge. For example, in cooperative positioning, the cooperators may offer misleading data owing to attacks or selfishness, affecting positioning accuracy. A blockchain-based secure and efficient GPS positioning error evolution sharing system can be established to improve vehicle location accuracy while ensuring the confidentiality and credibility of cooperators and their data. The GPS error sharing structure leveraging vehicular blockchain is shown in Figure 6.5.

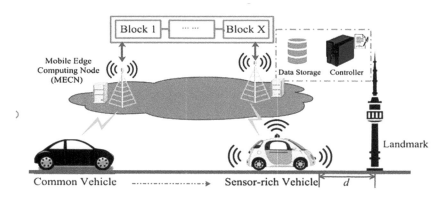

FIGURE 6.5
System architecture of the GPS error sharing framework using vehicular blockchain.

First, by analyzing GPS error, a bridge may be built between sensor-rich and sensor-less vehicles to foster cooperation by communicating the evolution of positioning error at a given time and location. Using an edge server-based deep neural network (DNN) prediction technique, blockchain-based positioning error storing and sharing ensures the safety of cooperative vehicles and Mobile Edge Computer Nodes (MECNs). The smart contracts are also set up to automate storage and tasks, as well as fix time scale inconsistencies.

Once the positional inaccuracy is calculated and the evolution of the present road network is obtained, the information can be shared with other vehicles in the smart city. A vehicle's GPS location can be obtained. Other models are employed to fix errors. This data is then stored by other network nodes. However, the cars may offer inaccurate data, and the processing nodes may be attacked, making them unreliable. So blockchain is important for safe data storage and sharing. [10] proposes a novel blockchain-based architecture based on Mobile Edge Computing Nodes (MECNs) on the blockchain network.

6.6.1 Blockchain in Information Sharing for Transportation Systems

Pre-selected nodes govern the consensus process in this method. The sensor-rich automobiles supply data, while the typical vehicles request it. Hence, the data provider just encrypts and sends the data as a blockchain transaction to adjacent MECNs. The mechanisms are required for data storage and smart contracts for data sharing.

Smart contracts assist reduce costs and protect the network from harmful behavior. The data supplier must first store the encrypted data in the system. An InterPlanetary File System utilized for content-addressed storage. Next, the data is published to the blockchain network. A message is issued to record one vehicle vi to VM.

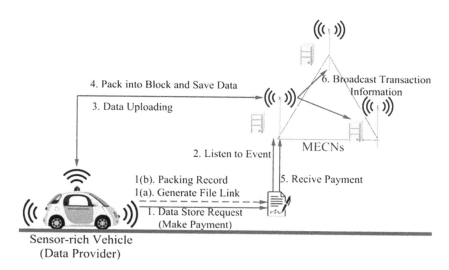

FIGURE 6.6
Smart contract process for data storage.

vi sent a message to a nearby computer node (Nm) with a unique identity (kipu), hash of the last updated Block h (Block-1), file link File Linki, and timestamp. The request message contains the identity kipu, the hash of the latest in Block h (Block-1), the file link File Linki, and the timestamp t (Figure 6.6).

Data is requested from the blockchain by a common vehicle, which picks the data to be sent based on the situation. The smart contract is activated by the information request. A request message is sent to the nearby node (Nm) if $(v\ j)$ selects vi's information. After sending the request, Nm verifies the certificate and checks on the IPFS for the requested data. This data is then stored as single-access linking, requiring single-access authentication from the data requester to ensure its security. The link becomes invalid after the data provider downloads it. The smart contract then pays (vi).

6.6.2 Blockchain in Vehicular Consensus Processing

The consensus process is broken down in the following procedure:

Broadcast: (vi) sends a request to MEC Nm, which acts as the consensus nodes. Nm then broadcasts the transactions to other nodes.

Pre-prepare: The leading computing node broadcasts a message. This then validates the transaction and then includes the validation result to the whole network.

Prepare: Nodes known as replicas receive the messages and transmit them on to other nodes on the network. The prepare message is broadcast to other networks after receiving the broadcast message from the leading or primary node.

Commit all consensus algorithms: Computing nodes commit messages to other nodes. A Replica checks the Prepare message and compares it to the leading node's pre-prepare message. The prepare message enters the commit state when it receives feedback from 2f replicas. This is the replication broadcast.

Reply (write block): Nm updates the blockchain. A replica will receive commit information. When it discovers that 2f + 1 (including itself) have authorized the information, it considers the transaction to be completed and attempts to notify the block producer. Each computing node receives data requests from numerous vehicles throughout time and records them.

6.7 VANET

With the widespread adoption of wireless communication [11] technologies, a world of connected things has emerged as seen in Figure 6.7. The Internet of things (IoT) is a new computer paradigm that connects many items with communication and sensing capabilities to the Internet. These gadgets connect to the Internet via WiFi, ZigBee, LTE, Bluetooth, and 5G. These technologies improve smart homes, smart cities, smart grid, etc. [12, 13]. In the near

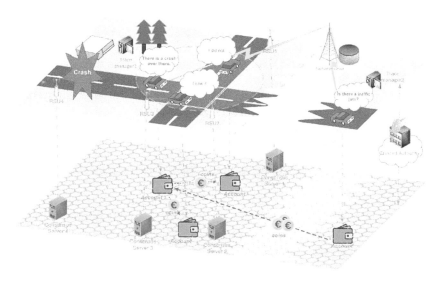

FIGURE 6.7
Vehicular blockchain consensus data processing.

future, billions of gadgets will be connected to the Internet, most of them cars. VANET connects millions of automobiles to increase traffic efficiency, which is a key component of IoT and smart cities. A VANET's peers (mainly automobiles) communicate by sharing road-related data, improving passenger and street safety, and routing traffic through congested areas.

Traffic view systems, message exchange, impact shirking, and secure accident reporting have all been developed extensively [14–16]. In VANET, cars have onboard sensors that intelligently exchange messages with other connected peers via V2V and V2I connections. 5G communication promises to be a strong technology that permits safe, rapid, and reliable vehicle-to-everything (V2X) connections. With high mobility, dynamic network topology, and data volume, 5G communication technology is predicted to meet future Internet of vehicles (IoV) application needs [17].

Due to the sensitive nature of VANET, it must always be a safe, trustworthy, and attack-free environment. Because every piece of information is vital, it must be authentic and provided by legitimate sources. Also, the network must meet security criteria [18, 19]. Aside from the high assurances and capacity VANET delivers, there is a need to properly manage, regulate, and run the system. Software Defined Networking (SDN) is thus offered [19] as a network technology capable of supporting dynamic VANETs and intelligent applications while cutting operational costs through simpler hardware, software, and management. The combination of SDN with IoV created Software Defined IoV (SD-IoV) [20], a technology designed to address issues in VANETs such as detecting efficient network heterogeneity exploitation and achieving different Quality of Service (QoS) needs.

Many implementations have focused on reliable message delivery between connected entities. As a result, verifying the quality of information given by peers has been neglected in order to protect the network from malicious peers. A centralized authority oversees the system's whole operation, posing security and privacy issues. Statuses of connected things must have girth descriptions. Severe privacy difficulties exist if the controller cannot be trusted or if the database is breached. Because the SDN is centralized, it is vulnerable to Denial of Service (DoS) assaults. It also has a single point of failure (SPF), thus a failing SDN controller response limits network connectivity.

A large number of false requests can disable the control capabilities. However, SDN and blockchain are used in VANETs to effectively manage and regulate the network. The blockchain, a distributed system, eliminates a centralized system's SPF. To manage incorrect information in VANETs for fog computing and 5G communication networks, the blockchain includes a trust-based message propagation module.

6.7.1 The Internet of Vehicles (IoV)

All participating vehicles in a typical VANET system are converted into wireless routers or mobile nodes. When a car in a network goes out of range or

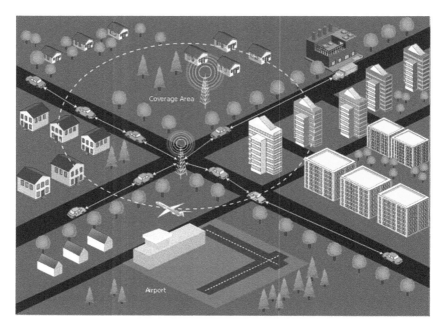

FIGURE 6.8
A simplified IoV environment.

drops out, other vehicles outside the network can always join. The network's complexity also makes it difficult to use. Because most objects are unpredictable and unstable, transient, and have a local or discrete use, VANETs are unable to provide global and durable applications for their customers. Vehicle intelligence and vehicle networking are two of the IoV's technical directions. The intelligence of the vehicle combines the intellect of the vehicle and the driver into a single entity.

As a result, the IoV focuses on the integration of human, vehicle, object, and environmental intelligence, and it is part of a wider network that provides services to cities. Multiple users, cars, items, and networks are all part of the IoV, which provides excellent manageability, operation, and control. As a critical component of smart cities, the purpose of IoV is to deliver high efficiency in transportation, improve services, lower costs, and realize in-depth integration of human-vehicle-object-environment. A simplified IoV structure is shown in Figure 6.8.

6.7.2 Software Defined Networking (SDN)

The SDN is a programmable network that revolutionizes network design and management. It turns network nodes into mere data/packet forwarding nodes by removing control functions. It also shifts data and route forwarding decisions from destination to flow-based. The SDN controller is an external entity in charge of the network's operations. A centralized controller with complete network knowledge can manage the entire network. The controllers communicate with the controller agent in the physical network devices via OpenFlow [21], an industry-standard protocol that is widely used.

6.8 Blockchain-SDN IoV Design

The planned architecture is shown in Figure 6.9. The frequent handover should be avoided due to vehicle movement and huge wireless communication between RSUs and cars. The fog computing platform is set up at the network's edge to overcome this issue. Infrastructural components of the platform include RSUs, base stations, and vehicles. The network's edges process and preserve most of the data sent between them. They are end-users since they have SDN-enabled on-board units. In addition to packet forwarding, On-Board Units (OBU) can collect environmental data and vehicle data such as speed and direction information.

They govern power, transmission mode, and channel. Moreover, the SDN controller controls the devices in the fog zones. The Road Side Units (RSU) hubs maintain vehicle control overhead in fog zones also known as Road Side Units Hubs (RSUHs). This is only done if asked by the SDN controller.

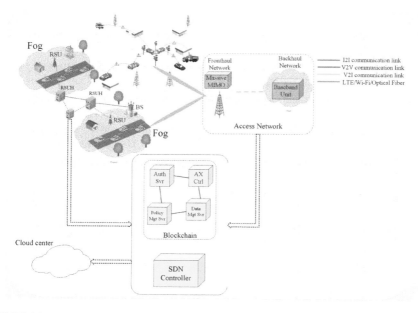

FIGURE 6.9
Blockchain-SDN IoV architecture.

Thus, RSUHs help reduce network overhead. They link several fog zones. RSUHs connect the VANET infrastructure to the SDN controller. They gather status data from connected fog cells and make forwarding decisions based on their own local knowledge. So the centralized controller has less overhead. The RSUHs and BBUs establish inter-zonal connection. RSUHs thus have data and control functions.

The SDN controller is responsible for resource allocation, mobility management, and rule development. The controller can also handle advanced network functions, including data pre-processing and analysis. The RSUs share the SDN's functionality hierarchically. The data plane, control plane, and application plane make up the logical structure of SDN. The data plane consists of vehicles, base stations, and RSUs and collects, quantifies, and forwards data to the control plane. The control plane which sets network flow rules includes the SDN controller, RSUHs, and a blockchain. The SDN controller will not completely control the network. However, the RSUHs and blockchain nodes share the tasks. The RSUHs will provide a network map based on data plane information.

The SDN model is made up of forwarding devices, the controller, and network applications. The forwarding devices receive packets, take actions on these packets, and update counters. Dropping packets, changing packet headers, and sending packets out of a single or several ports are types of the actions. The SDN controller gives instructions on how to handle packets. These devices will either know what to do with the packet or will question the controller. The SDN controller's network applications decide what to do with the packet and send information to the forwarding devices. The forwarding devices will then operate on the packet using the SDN controller as the translator. On the other hand, the forwarding devices will cache instructions so future packets don't need to be checked with the temporary forwarding device flow entries.

SDN can accommodate topological changes in vehicle networks with less software, hardware, and administrative expenses. This greatly helps IoT services in a VANET context. As a result of bandwidth, scalability, and connectivity limitations, gNodeBs are used in VANETs to provide wireless broadband Internet access. The SDN controller also controls RSUs and gNodeBs using high-capacity fiber optic backhaul lines. This helps with policy coherence and network management.

The 5G access network design is based on either the environment and/or the coverage. A wide range of environmental scenarios and coverage options are available. 5G will be deployed in densely populated areas in smart cities. This location frequently has huge cells and daily coverage. Dense population means high traffic loads and outdoor-to-indoor coverage. Therefore, the network will be complex, and new features will be expected to boost network efficiency. Technologies like Cloud and fog computing, SDN etc. will boost efficiency, while blockchain will offer a trustworthy and secure communication environment.

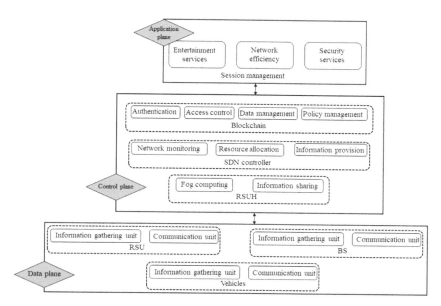

FIGURE 6.10
Network's logical view.

The blockchain is a decentralized database that maintains an immutable distributed ledger. The blockchain is used in this wireless network to cut administrative costs for spectral efficiency. Blockchain is used to secure network orchestration and, more critically, to assure message trust among cars. It maintains a distributed ledger to promote security and privacy as seen in Figure 6.10 in resource sharing, energy trade, etc. Because mining on the public blockchain incurs enormous expenses and delays, a consortium blockchain is deployed. A consortium blockchain can create a secure environment for data sharing by providing high security.

A blockchain is made up of three parts: transactions, blocks, and a consensus mechanism. The transaction includes data types, metadata, an encrypted link to the blocks and a timestamp. Digital signatures are used to guarantee the authentication of encrypted transactions. After signing, the transactions are stored in a cryptographically secure structure called a block, which is connected chronologically by hash references to form the blockchain. Then a consensus algorithm is devised to generate agreement on block order and validation.

The Practical Byzantine Fault Tolerance (PBFT) consensus algorithm is adopted. The RSUHs act as block miners in this case. A leader is chosen and a block is formed. After receiving the block, a limited set of pre-selected nodes vote to reach consensus. Rather, each node in a public blockchain must first solve a Proof of Work (PoW) puzzle before voting. PBFT is used to check a block's correctness. Due to its decentralized nature, each RSUH has full

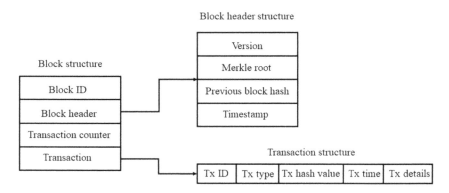

FIGURE 6.11
Block structure.

access to all transactions and can become a leader. In a consortium block-chain, the leader is picked before block generation and does not change until consensus. Figure 6.11 shows a block's structure.

- *Block structure* consists of the block ID, which indicates the block height.
- *The block header,* which has the block version, is used to keep track of software upgrades.
- *Previous block hash* references the hash of the previous (parent) block.
- *Merkle root* that is a hash of the root of the Merkle tree of the block's transactions.
- *Timestamp* specifying the time the block was created.
- *The transaction counter* keeps track of the number of transactions in the chain.
- *Transaction field* specifies the transactions recorded in the chain. Similarly, the transaction field consists of several elements such as the
  1. *Transaction ID* uniquely identifies a particular transaction in the chain,
  2. *Transaction type* which determines the kind of transaction being initiated,
  3. *Transaction hash* is the hash value generated from that particular transaction,
  4. *Transaction time* determines the time the transaction was made, and finally,
  5. *Transaction details* that provides a brief description of the transaction.

An authentication server, an access controller, a data management server, and a policy management server make up the blockchain nodes. To provide effective and efficient network administration, the blockchain works in tandem with the SDN.

- *Authentication server*: Authentication services are provided by the authentication server to the network's entities. An entity must be registered before becoming a member of the network, and cryptographic keys must be issued to that entity. The entity must be authenticated after successful registration and before establishing any type of network connection. The server authenticates authorized network users to avoid malicious activity, as a network hack would result in serious security and privacy concerns.

- *Data management server*: The data management server assumes some of the SDN controller's management responsibilities. The network peers will have access to and process multiple requests in such a large, broad environment. Because the SDN is centralized, there is a risk of a SPF, which will significantly slow down the network's performance. The blockchain's decentralized structure assures that the SPF problem is solved. When the SDN controller can't handle all of the incoming requests, it sends some of them to the data management services, where the necessary measures can be taken.

- *Access control*: Every network has an access control policy to ensure that the data generated is only used by the appropriate entities. The access controller is in charge of this. Before data is sent to a network peer, it must first be checked to see if the entity is part of the data owner's access list. The access controller provides the entity access to the data if the check is successful. They are dropped if the request is unsuccessful.

- *Policy management server*: Lastly, a policy management server penalizes the network's defaulting members. The smart contract, which is a programmable script that invokes activities based on given rules in the contract, is a critical component of the blockchain. When a vehicle delivers misleading information that leads to an accident in the network, the policy management server penalizes the car, which can take the form of toll tickets. The introduction of the trust model in the network was motivated by such false information.

Secure and intelligent resource management, as well as flexible networking, are among the system's benefits. To aid in safe resource sharing, the blockchain provides tamper-resistant block production. SDN can recognize complex wireless networks and the varying needs of developing services, allowing for the most efficient resource allocation policy in a secure communication environment. Furthermore, the presence of diverse network components allows for flexible networking. SDN may analyze the network's topology, channel assignment, and packet forwarding, and then choose the best wireless modes to improve communication, reduce energy consumption, and improve user experience. Finally, the use of blockchain allows for decentralized synchronization and replication of network configurations, which will make network orchestration and diagnosis easier. Furthermore, computations for blockchain-enabled systems become simpler.

6.9 Certificate Issuance and Revocation

Certificate issuance and revocation are addressed, as well as two communication and trust models, all of which are put to the test to see how successful and efficient they are.

6.9.1 Certificate Issuance and Revocation

As long as the private keys stay intact, cryptographic algorithms of the Public Key Infrastructure (PKI) enable secure communication between entities in a network. This is the responsibility of the authentication server, which ensures the authenticity and integrity of delivered messages. This unit also maintains a database with a list of vehicles' public keys and identities in order to protect the vehicles' anonymity and ensure their legality. In addition, the unit can track vehicles in the event of a dispute in order to acquire evidence.

The blockchain has hardware that makes it an ideal choice for high-performance computations, such as data processing, storage, and transmission, which are limited in RSUs and automobiles. Each broadcast vehicular message includes a timestamp, the sender's address, and digital signatures from the authentication server and the certificate issuance authority (CIA). The CIA is a part of the authentication server that is in charge of granting certificates and revoking public keys under the authentication server's supervision. The date of expiry and the approved public key are both included in a certificate transaction. The revoked public key is contained in a revoked transaction.

Each entity generates a public and private key pair $\{P_{pub}, P_{priv}\}$. Vehicle A enters the network and transmits its public key and other identification documents to the authentication server via a secure connection to establish its identity. If the materials provided are valid, the server will submit a signed clearance to the CIA. The CIA then gives the car with a certificate. Vehicle A's contents are only accessible to the authentication server since they contain private information that will be used to monitor the vehicle's identity in the event of a disagreement.

The following are the contents of Vehicle A's certificate: tuple $C_A = \{P_{pub_{VA}}, P_{pub_{CIA}}, P_{pub_{AS}}, X_{CIA}, X_{AS}, RS_{C_{VA}}, ExT_{VA}\}$, where X is the signature, RS_C stands for the vehicle's reputation score, AS stands for the authentication server, and ExT stands for the certificate's expiration period. Vehicle A produces its public and private key pair $\{P_{pub_{VA}}, P_{priv_{VA}}\}$ and makes a certificate request to the authentication server to receive a certificate. This request contains $P_{pub_{VA}}$ and the signature X_{VA}, and it will be encrypted with the server's public key $P_{pub_{AS}}$. After a successful verification process, the authentication server will send this signed clearance to the CIA. The CIA issues a certificate, including the vehicle's reputation score, $RS_{C_{VA}}$, and the period until it expires, ExT_{VA} after checking and validating the signature in the clearance. The blockchain keeps track of this data.

The authentication server transmits a signed revocation warrant to the CIA for a revoked user owing to malicious conduct, which includes the revoked public key, $P_{pub_{rvk}}$, and the time of revocation, T_{rvk}. The CIA will then broadcast this transaction, which includes the timestamp, the revoked public key, and the signatures of both the CIA and the authentication server.

The revocation transaction will be verified by all associated entities, and their databases will be updated. The following tuple makes up a revoked transaction: $RV = \{P_{pub_{rvk}}, P_{pub_{CIA}}, P_{pub_{AS}}, X_{CIA}, X_{AS}, T_{rvk}\}$.

6.9.2 Communication Model

Mobile ad hoc networks (MANETs) on the road are made up of Dedicated Short Range Communication (DSRC) devices (IEEE 802.11p) [22] that are installed in vehicles. These improve intelligent transportation system (ITS) applications in situations when communication latency and reliability are critical. IEEE 802.11p, also known as Wireless Access in Vehicular Networks (WAVE), is a 5.9 GHz frequency band standard that uses Orthogonal Frequency Division Multiplexing (OFDM) to provide V2V and V2I connections up to 1000 m. The WAVE spectrum is broken into channels with numbers ranging from 172 to 184 and a 2-second gap between them. The control channel (CCH) is where all of the devices listen for information and where safety messages are communicated. OFDM has a 75 MHz bandwidth with up to seven 10 MHz channels and a 5 MHz guard band. The first (CH 172) and last (CH 184) channels are both unused, with the last (CH 184) designated for future usage (High Availability Low Latency [HALL] channel). The remaining channels are service channels (SCHs), which are used to broadcast IP-based services.

In each communication zone, the CCH and at least one of the four SCHs are used for communication. Each channel has a 10 MHz frequency bandwidth, as opposed to the 20 MHz used in the IEEE 802.11a standard, to increase its tolerance to multipath propagation and the effects of Doppler dispersion on the network. Because of its interference resistance and low coding rate (about 0.5), Binary Phase-Shift Keying (BPSK) is also used [23]. The signals are sent at a 3 Mbps data rate, with each packet sized at 500 bytes, which is the average number found in various security studies [24]. In addition, a packet can only be correctly received if it arrives after 10_μ on the reception of the previous packet. Requiring a minimal Signal-to-Noise ratio (SNR), an SNR value of 5 dB is set.

Furthermore, the probabilistic Nakagami distribution [25] is used because it is well-suited to estimating fading events in mobile communication channels. In the Nakagami distribution, the probability of receiving a signal decreases as the distance increases. From Rayleigh to Additive White Gaussian Noise (AWGN) channels, Nakagami covers it all. The expression represents the probability density function (pdf) of the Nakagami-m fading.

$$f_v(v) = \left(\frac{m}{\bar{v}} \right)^m \frac{v^{m-1}}{\Gamma(m)} \exp\left(m\frac{v}{\bar{v}} \right), m \geq \frac{1}{2} \tag{6.1}$$

where v is the received signal and $\bar{v}$ is the average received SNR. m is the Nakagami fading figure, and $\Gamma(\cdot)$ is the Euler gamma function, represented as $\Gamma(a) = (\alpha - 1)!$. The Nakagami fading model has been shown to accurately represent a variety of indoor and outdoor multipath mobile signals. The channel approaches an AWGN channel as m approaches infinity. The Nakagami model is utilized in this study to assess the fading channel's capacity and to determine which ranges have a high probability of message reception, on which to base the experiment. The expression is what distinguishes m.

$$m \triangleq \frac{\Omega^2}{E\left[V^2 - \Omega^2 \right]} \tag{6.2}$$

where $\Omega = E\left[V^2 \right]$.

Given an average transmit power, the capacity of the fading channel with the pdf function is given by

$$C = B \int_{v_0}^{+\infty} \log_2 \left(\frac{v}{v_0} \right) f_v(v) dv \tag{6.3}$$

where B is the bandwidth of the channel in Hertz (Hz), and v_0 is the cutoff level of the SNR, below which there is a suspension in data transmission. Optimal cutoff should satisfy

$$\int_{v_0}^{+\infty} \left(\frac{v - v_0}{vv_0}\right) f_v(v)dv = 1 \tag{6.4}$$

In order to obtain maximum capacity (Equation (6.3)), the channel fade should be tracked at both the transmitter and the receiver, as well as rate and power adaptation at the transmitter. For good channel conditions (a big v), a high power allocation should be provided, and vice versa. The probability of no transmission is given by

$$P_{nt} = \int_0^{v_0} f_v(v)dv = 1 - \int_{v_0}^{+\infty} f_v(v)dv \tag{6.5}$$

stemming from the fact that at a level where $v < v_0$, there is no data sent.

Putting Equation (6.1) into Equation (6.4), v_0 must meet the condition

$$\frac{\bar{v}\Gamma\left(m, m\frac{v_0}{\bar{v}}\right)}{v_0} - m\Gamma\left(m-1, m\frac{v_0}{\bar{v}}\right) = \bar{v}\Gamma(m) \tag{6.6}$$

When $m = 1$, the Nakagami distribution reduces to Rayleigh fading (suitable for non-line-of-sight [nLOS] wireless channels). Equation (6.6) therefore reduces to

$$\frac{\bar{v}e^{-v_0/\bar{v}}}{v_0} - E_1\left(\frac{v_0}{\bar{v}}\right) = \bar{v} \tag{6.7}$$

where $E_1(.)$ represents the first order exponential integral.

Defining the integral ζ_r as

$$\zeta_r(\varphi) = \int_1^{+\infty} u^{r-1} \ln(u) e^{-\varphi u} du \tag{6.8}$$

and substituting Equation (6.1) into Equation (6.3), the channel capacity can be rewritten as

$$C = \frac{B\log_2 e}{\Gamma(m)} (m\kappa)^m \zeta_r(m\kappa) \tag{6.9}$$

where $\kappa = v_0/\bar{v}$. From this, the channel capacity per unit bandwidth can be deduced as

$$\frac{C}{B} = \log_2(e) \sum_{i=0}^{m-1} \frac{\Gamma(i, m\kappa)}{i!} \tag{6.10}$$

This can also be written as

$$\frac{C}{B} \log_2 (e) \left(E_1 (m\kappa) + \sum_{i=0}^{m-1} \frac{P_i (m\kappa)}{i} \right) \tag{6.11}$$

where $P_i(.)$ is a Poisson distribution, given by

$$P_i (\varphi) = e^{-\varphi} \sum_{j=0}^{i-1} \frac{\varphi^j}{j!} \tag{6.12}$$

Using the Rayleigh fading channel, and from Equation (6.7), the channel capacity becomes

$$C = B \log_2 (e) \left(\frac{e^{-\kappa}}{\kappa} - \bar{v} \right) \tag{6.13}$$

The probability of no transmission from Equation (6.5) can be rewritten using Equation (6.1) as

$$P_{nt} = 1 - \frac{\Gamma(m, m\kappa)}{\Gamma(m)} = 1 - P_m (m\kappa) \tag{6.14}$$

When no fading is taken into account, the likelihood of a successful message reception in the communication range studied is less than 1. In addition, as demonstrated in Figure 6.6, as the fading figure increases, the probability of signal reception decreases. Gruff, medium, and moderate fading figures are referred to as 12, $m = 3, 5,$ and 7, respectively. In wireless contexts, however, OFDM receivers alleviate temporal and frequency selectivity issues. The graphic shows that within the ranges of 200 m and 500 m, there is a high possibility of message receipt. There is a probability of a successful reception even if you're beyond the 500 m communication range (less than 0.3) as shown in Figure 6.12.

The simulation scenario consists of an 8-km long highway that is bi-directional and has four lanes in each direction. Averagely, there are ten vehicles per kilometer in each lane and an average traveling speed of 80 kmph. All simulations are carried out in MATLAB and NS-3 simulator [26].

6.9.3 Network Trust Model

Unlike MANETs, VANETs face various obstacles that make it difficult to design an effective trust model, including high mobility, frequent topological changes in the network, and the ability to eavesdrop on broadcast

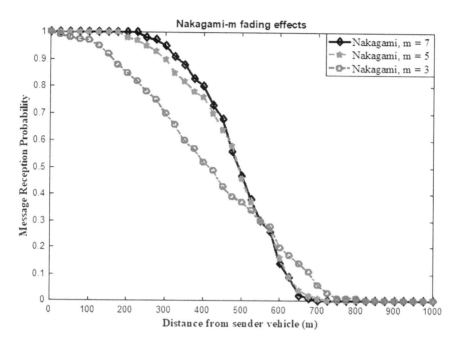

FIGURE 6.12
Probability of message reception against distance.

messages from vehicles. Due to the limitations outlined earlier, cars may only exchange a smattering of messages, making long-term vehicle contact unfeasible. In the context of congestion and sparse environments, a trust model should be built due to network topology changes. Cooperative information from numerous vehicles and RSUs is required to validate the integrity of the received data, and this information must be immutable, reliable, and authentic. Messages from multiple senders, as well as all verdicts on the message carried by other vehicles, should be stored as evidence in the blockchain. The blockchain can be inferred in circumstances where a vehicle's reputation is merited. This aids in the active and open interchange of information across the entities.

A trusted network is used to verify that the information provided by peers (vehicles) is reliable enough to be followed. Peers in the network must also give their opinions on each piece of information (deemed helpful) provided by the vehicles. The leader of a clustered vehicular network evaluates the various verdicts of the information supplied by the sender in order to detect faked information provided by the peer, and if the weight is above a threshold (also determined by the leader), a judgment on the message's trustworthiness is taken. In the network, we propose the following types of communications: a sender message sent by a vehicle, a verdict on the message's trustworthiness, and finally, the sum total message. The event being reported, the sender's confidence in the message, the time of the event reporting, and the place where the event occurred should all be included in the sender's message. When it comes to the verdict of the other vehicles on the message, the vehicles should decide whether they believe the message and submit their response to it. A person's response to a communication is to either trust or distrust the reported event. The sender's message is combined with the numerous judgments offered by the connected peers to create the total sum message.

The model is divided into three algorithms, each of which is described in the following:

1. *Cluster model*: In the cluster model, cars are spatially divided into clusters. Because messages are relayed between cluster leaders rather than two neighboring peers, clustering improves network scalability. Each cluster selects one vehicle to serve as the group's leader, L. When a peer, say R, sends a message, M, to the group, the members must give the message a verdict, V. As an example, a message from a vehicle, say, R_i, will be read as $M_0 = [M, ID_0, X_0]$, where ID_0 is the ID of the sender, and X_0 is the signature on the message M_0. Each vehicle in the cluster will provide a verdict on the message. The cluster leader then takes the various verdicts of the members, $M_i = [M, V_i, ID_i, X_i]$, and provides a sum total message M_{ST}, i.e., $M_{ST} = [M, V_i, \ldots, V_\beta, ID_i, \ldots, ID_\beta, X^*], i \in [1, \beta]$, where X^* is the aggregation of all the signatures.

2. *Message forwarding model*: After all of the verdicts on a message have been compiled, the message must be transmitted to other clusters in the network to raise awareness. The message must be forwarded to other peers based on its trustworthiness [27], or it will be dropped. The communication is transmitted to other clusters when the majority of the cars in the network trust it; otherwise, it is dropped. When a message is broadcast across the network, the set of peers who trust the message is denoted by R, while those who do not trust the message are denoted by R'. The weight of W of the trust and distrust outcomes is then calculated by the cluster leader. It then sets a threshold, α, with which the weight is compared. If $W \geq \alpha$, the message is forwarded. If $W < \alpha$, the message is dropped.

3. *Judgment model*: The critical model is concerned with decisions made after many verdicts have been received from a sender's message. The network peers select whether or not to conform to the information in the message based on the aggregated trustworthiness represented as TA over a message M.

6.10 Conclusion

This chapter looked at smart transportation for smart cities. We looked at ways of ensuring a highly efficient networks for transport system and also technologies built to ensure safety over transport networks. We take a look at MaaS and the solution provided for transport operators and data providers to share transport information. We also look at the security challenges in securing big data for mobility. We examine some solutions toward ensuring effective vehicular announcement networks. Finally, we observe a blockchain-based information sharing protocol and consensus processes for VANETS, making use of the decentralized and immutable characteristics of blockchain network.

References

[1] I. Yaqoob, I. A. T. Hashem, Y. Mehmood, A. Gani, S. Mokhtar, and S. Guizani, "Enabling Communication Technologies for Smart Cities," in IEEE Communications Magazine, vol. 55, no. 1, pp. 112–120, Jan. 2017.

[2] S. Chen, H. Xu, D. Liu, B. Hu, and H. Wang, "A Vision of IoT: Applications, Challenges, and Opportunities with China Perspective," in IEEE Internet of Things Journal, vol. 1, no. 4, pp. 349–359, Aug. 2014.

[3] Y. Mehmood, F. Ahmad, I. Yaqoob, A. Adnane, M. Imran, and S. Guizani, "Internet-of-Things-Based Smart Cities: Recent Advances and Challenges," in IEEE Communications Magazine, vol. 55, no. 9, pp. 16–24, Sept. 2017.

[4] T. Nadeem, S. Dashtinezhad, C. Liao, and L. Iftode, "Trafficview: Traffic data dissemination using car-to-car communication," ACM SIGMOBILE Mobile Computing and Communications Review, vol. 8, no. 3, pp. 6–19, 2004.

[5] Q. Xu, T. Mak, J. Ko, and R. Sengupta, "Vehicle-to-Vehicle Safety Messaging in DSRC," in Proceedings of the ACM International Workshop on Vehicular Ad Hoc Networks, pp. 19–28, 2004.

[6] T. ElBatt, S. K. Goel, G. Holland, H. Krishnan, and J. Parikh, "Cooperative Collision Warning Using Dedicated Short Range Wireless Communications," in Proceedings of the ACM International Workshop on Vehicular Ad Hoc Networks, pp. 1–9, 2006.

[7] S. U. Rahman and U. Hengartner, "Secure Crash Reporting in Vehicular Ad Hoc Networks," in 2007 Third International Conference on Security and Privacy in Communications Networks and the Workshops – SecureComm 2007, Nice, France, pp. 443–452, 2007.

[8] X. Ge, Z. Li, and S. Li, "5G Software Defined Vehicular Networks," in IEEE Communications Magazine, vol. 55, no. 7, pp. 87–93, Jul. 2017.

[9] H. Zhong, B. Huang, J. Cui, Y. Xu, and L. Liu, "Conditional Privacy-Preserving Authentication Using Registration List in Vehicular Ad Hoc Networks," in IEEE Access, vol. 6, pp. 2241–2250, 2018.

[10] Y. Sun, L. Wu, S. Wu, S. Li, T Zhang, L. Zhang, J. Xu, and Y. Xiong, "Security and Privacy in the Internet of Vehicles," in Proceedings of the International Conference on Identification, Information, and Knowledge in the Internet of Things (IIKI), Beijing, China, pp. 116–121, October 22–23, 2015.

[11] M. Batty, K. W. Axhausen, F. Giannotti, A. Pozdnoukhov, A. Bazzani, M. Wachowicz, ... and Y. Portugali, "Smart Cities of the Future," in The European Physical Journal Special Topics, vol. 214, no. 1, pp. 481–518, 2012.

[12] Z. Lu, G. Qu, and Z. Liu, "A Survey on Recent Advances in Vehicular Network Security, Trust, and Privacy," in IEEE Transactions on Intelligent Transportation Systems, vol. 20, no. 2, pp. 760–776, Feb. 2019.

[13] J. Cui, W. Xu, H. Zhong, J. Zhang, Y. Xu, and L. Liu, "Privacy-Preserving Authentication Using a Double Pseudonym for Internet of Vehicles," in Sensors, vol. 18, p. 1453, 2018. 10.3390/s18051453.

[14] S. Sezer et al., "Are We Ready for SDN? Implementation Challenges for Software-Defined Networks," in IEEE Communications Magazine, vol. 51, no. 7, pp. 36–43, Jul. 2013.

[15] K. Zheng, Q. Zheng, P. Chatzimisios, W. Xiang, and Y. Zhou, "Heterogeneous Vehicular Networking: A Survey on Architecture, Challenges, and Solutions," in IEEE Communications Surveys and Tutorials, vol. 17, no. 4, pp. 2377–2396, Fourth quarter, 2015.

[16] J. Chen, H. Zhou, N. Zhang, P. Yang, L. Gui, and X. Shen, "Software Defined Internet of Vehicles: Architecture, Challenges and Solutions," Journal of Communications and Information Networks, vol. 1, pp. 14–26, 2016.

[17] G. Araniti, C. Campolo, M. Condoluci, A. Iera, and A. Molinaro, "LTE for Vehicular Networking: A Survey," in IEEE Communications Magazine, vol. 51, no. 5, pp. 148–157, May 2013.

[18] P. K. Sharma, S. Y. Moon, and J. H. Park, "Block-VN: A Distributed Blockchain Based Vehicular Network Architecture in Smart City," JIPS, vol. 13, no. 1, pp. 184–195, 2017.

[19] L. Xie, Y. Ding, H. Yang, and X. Wang, "Blockchain-Based Secure and Trustworthy Internet of Things in SDN-Enabled 5G-VANETs," in IEEE Access, vol. 7, pp. 56656–56666, 2019.

[20] N. McKeown, T. Anderson, H. Balakrishnan, et al. "OpenFlow: Enabling Innovation in Campus Networks," ACM SIGCOMM Computer Communication Review, vol. 38, no. 2, pp. 69–74, 2008.

[21] M. Al-Bassam, "SCPKI: A Smart Contract-Based PKI and Identity System," in Proceedings of the ACM Workshop Blockchain, Cryptocurrencies Contracts, Association for Computing Machinery, New York, NY, pp. 35–40, 2017

[22] IEEE, "IEEE P802.11p/D5.0, Draft Amendments for Wireless Access in Vehicular Environments (WAVE)," Apr. 2009. http://ieeexplore.ieee.org/servlet/opac?punumber=4810961.

[23] J. Maurer, T. Fügen, and W. Wiesbeck, "Physical-Layer Simulations of IEEE802.11a for Vehicle-to-Vehicle Communication," in Proceedings of the IEEE 62nd VTC—Fall, Dallas, TX, vol. 23, pp. 1849–1853, Sep. 2005.

[24] M. Raya and J. Hubaux, "The Security of Vehicular Ad Hoc Networks," in Proceedings of the 3rd ACM Workshop SASN, Alexandria, VA, pp. 11–21, Nov. 2005.

[25] M. Nakagami, "The m-Distribution: A General Formula of Intensity Distribution of the Rapid Fading," in Statistical Methods in Radio Wave Propagation, W. C. Hoffman, ed. Oxford, UK: Pergamon, 1960.

[26] R. K. Jaiswal and C. Jaidhar, "An Applicability of AODV and OLSR Protocols on IEEE 802.11p for City Road in VANET," in Internet of Things, Smart Spaces, and Next Generation Networks and Systems, Springer International Publishing, Switzerland, pp. 286–298, 2015.

[27] E. Androulaki, A. Barger, V. Bortnikov, C. Cachin, K. Christidis, A. De Caro, and S. Muralidharan, "Hyperledger Fabric: A Distributed Operating System for Permissioned Blockchains," in Proceedings of the Thirteenth EuroSys Conference, Association for Computing Machinery, New York, NY, p. 30, Apr. 2018.

7

Smart Health

7.1 Introduction

Private health data, electronic medical data, and electronic health records have all surfaced as valuable assets with the ability to influence the quality of life for people all over the world. The World Health Organization as an asset, the dissemination of which extends far beyond its immediate medical utility, has previously acknowledged personal health information [1, 2]. However, due to fragmentation and isolation of personal health data in hospitals and healthcare provider networks to comply with industry and government regulation, making full use of the data is difficult. The true power of [3] this crucial [4] asset is awakened when data is available, accessible, and complete at all times. The true power of this essential asset is unlocked when data is available, accessible, and complete at the point of care when it is required. Digital health information exchanges [5, 6] were an early solution [7–9] that enabled digital transfer of patient medical records between providers and institutions. To name a few, this has aided in the resolution of issues such as redundant testing and associated costs, improved coordination of patient care, lower total administrative expenses [10], and improving patient quality of life benefits.

Despite repeated interventions, a number of problems have persisted and defied attempts to solve them. Some of them include interoperability [11], data security and privacy [12, 13], trust, and data control following transmission to a requester. The data sought may be subjected to a variety of computations, over which existing approaches have little-to-no control. Legal and regulatory sanctions exist for the illegal or unethical use of requested information. On the other hand, these can only be employed after the crime has been committed and detected by the appropriate authorities. At this point, the reputational damage may be irrevocable. Despite the difficulties [14–16], the trend of linking disparate data sources, as witnessed in the healthcare industry, is unavoidable. Blockchain, the distributed ledger technology that supports the Bitcoin cryptocurrency, gives users control over the data they share and a way to track what they do with it after it has been shared. The effective and intelligent use of cryptographic [17, 18] primitives combined with a consensus

DOI: 10.1201/9781003289418-7

method provides a highly secure platform for the performance of computations with applications in numerous [19] health fields. Several studies have been conducted on health information sharing trends, with a focus on data control and security afforded by blockchain-based smart contracts. Despite the rapid expansion of blockchain-enabled health applications and research, few people pay attention to active data control and/or protection once it has been disseminated or shared.

Patientory is a Software-as-a-Service (SaaS) [20] platform that creates a unique profile for each patient to provide them secure access to their medical records. The medical data is stored on a blockchain platform that complies with Health Insurance Portability and Accountability Act of 1996 (HIPAA) regulations, guaranteeing that both patients and caregivers have access to confidential information. Patients' health plans can be viewed on Patientory by care providers. Another blockchain-based health information system architecture is the Health Data Gateway (HDG) that was introduced by Onik et al. [21]. Patients will be able to own, control, and share their medical data utilizing private blockchain Clouds, according to the project's goals. Xia Qi et al. [10] released Blockchain-based Data Sharing (BBDS), a secure medical data access and sharing framework for a pool of sensing devices. However, in the portion where they looked into the use of smart contracts and blockchain technology, the writers solely considered data provenance and auditing.

MeDShare's design [11], which is a natural extension of BBDS [10], integrates with existing trustless Cloud providers and other data guardians [22] that want to safely transfer and share data between entities. MedRec is yet another blockchain-based decentralized record management system. MedRec's [12] modular design manages authentication, secrecy, responsibility, and data exchange. Ancile, like MedRec, is a record management system that improves access control and obfuscates data using smart contracts and an Ethereum-based blockchain. Despite the huge potential of blockchain technology for enhancing medical data access, sharing, and permission management, there is currently a need in the literature for a data-centric study that focuses on data control after data is transmitted from one entity to the next in the network. Industry regulations must be followed when designing systems. The monitoring of system interactions should start when the user makes a data request and terminate when the system responds to the user. Then only such monitoring should be stopped.

This chapter offers a blockchain-based digital health information exchange with an underlying mechanism for monitoring and enforcing patient data's authorized usage restrictions. Patients in the system write policies at registration that define what activities they can do with their personal health data. These policies are saved in the system and used in conjunction with smart contracts to decide whether or not data can be shared. Participating health institutions' processing nodes, smart contracts, and security monitors work together to ensure patient data security is secured from unauthorized access and computations.

7.2 Preliminaries

This section describes the system elements that interact to provide the data control and protection service in the health information exchange. The function of the blockchain in the process is highlighted, as well as the blocks that were built to make it even more secure.

7.2.1 Cryptographic Keys

Our system relies on a set of cryptographic keys that are generated during the registration process and enable users to conduct specific tasks. These ensure that users who have been allowed access to the system can only make data requests. Data security is also ensured by our system. We connect the blockchain with an efficient public key cryptosystem currently in place, forcing entities to interact with the blockchain in order to obtain information. The user using parameters provided by the Registration Authority creates this key, and it is used to sign requests for data access. It is stored on the user's computer and is not shared with any other system component.

- *Transaction public key*: This key is generated by the user and sent to the blockchain network. The system requires the Query Manager to retrieve this key in order to verify the signature on a data request.
- *Membership key*: The membership key is generated by the Registration Authority when a new user requests to join a permissioned group. The Registration Authority generates a SHA256 hash of the key when it is handed to the new user and sends it to the Query Manager for storage. The membership key is attached to data requests before they are sent to the Query Manager. The Query Manager is an application that allows you to construct queries.

The user must first create a request and sign it with the private key generated during registration in order to gain access to system data. The Registration Authority's (RA) membership key is then added, and the request is passed to the Query Manager. The Query Manager verifies the request using the transaction public key obtained from the blockchain. The processing nodes, who then respond with a response, then, process the request.

7.2.2 Smart Contracts

Smart contracts are blockchain-based programs that all nodes on the network execute. They contain codes that outline the terms and conditions of mutually agreed-upon contracts. Their existence on the blockchain network ensures that they cannot be tampered with, offering the assurance of trust that could previously only be obtained through stringent controls and audit processes carried out by reputable third parties. Smart contracts make it possible to create dynamic and scalable rules that can be used to securely transfer patient data in a health information exchange. They are useful for reorienting operations that remove human actors from the system, increasing system efficiency. Smart contracts added to the health information exchange add another layer of security while also enhancing efficiency.

7.3 System Design

For the trading platform, the data control mechanism is constructed utilizing a blockchain-based method. The many entities used in our system are depicted in Figure 7.1. The following are the structural components of the system:

1. *Users*: This includes doctors and patients who need data access to carry out their responsibilities. During registration, users are allocated identities that help identify them as doctors, patients, and so on, as well as decide access privileges and actions that are approved on the data they seek based on policies that have been recorded. This includes doctors and patients who require data access in order to do their duties. Users are assigned identities during registration that assist in identifying them as doctors, patients, and so on and determine access rights and activities that are authorized on the data they seek based on recorded policies.

2. *Query manager*: A person in charge of queries is referred to as a query manager. The query manager receives queries and formats them in a standard fashion before sending them to the network. It first receives the data request policy and compares it to the request's objective.

3. *Smart contract center*: The smart contract center is a non-profit organization dedicated to promoting smart contract adoption. This entity collaborates with the processing nodes to create smart contracts that implement policies on saved patient data. It generates smart contracts and connects them to the data requester's request, as well as retrieving responses.

4. *Processing nodes*: Nodes that process data are known as processing nodes. This is a network of nodes that take data requests and process them while providing request validation to the query manager. Requests and responses are processed by processing nodes, which aid in the creation of blocks. They also put together packets with the answers to questions.

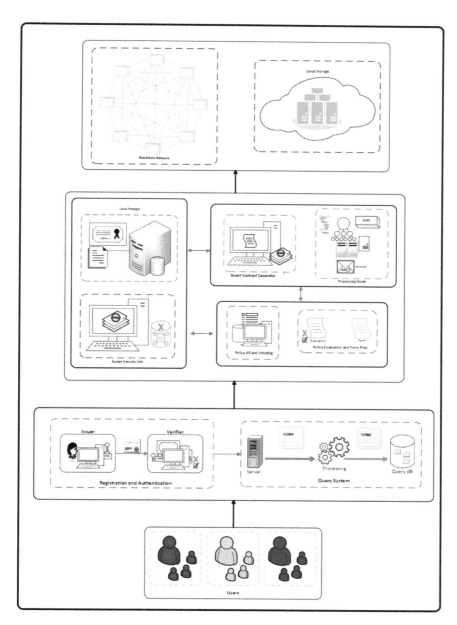

FIGURE 7.1
System architecture.

5. *Local storage*: The local storage acts as a cache for policies, tokens, and parameters that are utilized in the process of granting access to patient data.

6. *Blockchain network*: The term "blockchain network" refers to a network of computers that are linked together. They are received and stored by the blockchain network as transaction logs that have been processed into blocks. It records the details of the transaction, as well as the data's statuses, policies, and requests. A blockchain [23] is a distributed database that comprises a chronologically ordered set of information linked together by placing the previous block's hash on the next block for our needs.

7.4 Construction

We will go through how to create entities and component functions that allow you to control data operations after they've been disseminated in this section. We also go through the system architectures that help with data control and privacy on the platform.

1. *Registration*: A user must first register their credentials with the Registration Authority before setting up policies for data sharing and allowed use. The user approaches the Registration Authority with a request to join the permissioned group. Based on a number of variables, the RA generates a membership key for the user. Following that, it creates a SHA256 hash. A user must first register their credentials with the Registration Authority before setting up policies for data sharing and allowed use. The user approaches the Registration Authority with a request to join the permissioned group. Based on a number of variables, the RA generates a membership key for the user. The key is subsequently hashed and sent to the receiver as a SHA256 hash. On their platform-stored records, the user then generates a transaction key pair as well as a policy set for sharing and other activities. The transaction's public key is transferred to the blockchain network, and the policy set is kept on the storage layer. The user then provides a SHA256 hash of the membership key acquired from the RA to the query manager. This additional authentication establishes the user as a legitimate group member. The query manager sends an acknowledgment message to the Registration Authority to signal the completion of registration and to complete the procedure.

2. *Request*: The user signs a data request with the transaction private key and the membership key and sends it to the query manager.

The query manager receives the message and begins a two-step verification process. It begins by using SHA256 to hash the membership key. If the hash received from the Registration Authority when the user first registered matches, the request is considered legitimate. After that, the query manager retrieves the transaction public key from the blockchain and verifies the signature of the request. Once the signature verification is successful, the query manager retrieves the policies set created at sign on for the data. After that, the permissions are looked at to see what operations on the data are allowed. After this stage, the request is forwarded at time *t*. The request is then sent on to the processing nodes that connect with the smart contracts center and the security monitor to ensure that the user's data is being tracked.

3. *Processing/reply*: The processing nodes are in charge of handling requests from users of the system. They send data to the smart contracts center after retrieving it from the storage layer. The smart contracts center then creates and appends smart contracts to the request before delivering it to the processing nodes to be sent to the user. The methods for merging the data with the smart contracts and policy in the reply are not addressed in this writing because they are outside the scope of this study.

4. *Monitoring*: The specified instructions for monitoring how the data was used are included in the smart contract that came with the response. It continuously monitors the connection between the storage layer and the user node using a set of timing and connectivity functions. The connectivity between the two nodes is utilized to make monitoring the smart contracts' timing capabilities easier. The timers are detection timers for the system. The timers are set by the system and are based on a reasonable amount of time for data processing to be completed. Our approach assumes unlawful operations that are planned because monitoring is impossible when connectivity is lost. The timers are reset to zero when connectivity is lost, and the data is destroyed using smart contract instructions. The incident is recorded in a report as seen in Figure 7.2.

7.5 Conclusion

As part of the ongoing issues of health information exchanges, data control challenges have been identified as security and privacy concerns. We have provided a blockchain-based strategy for accessing data and monitoring its use in this chapter. The technique, which is based on smart contracts, monitors data after it is received by the requester and, if violations occur

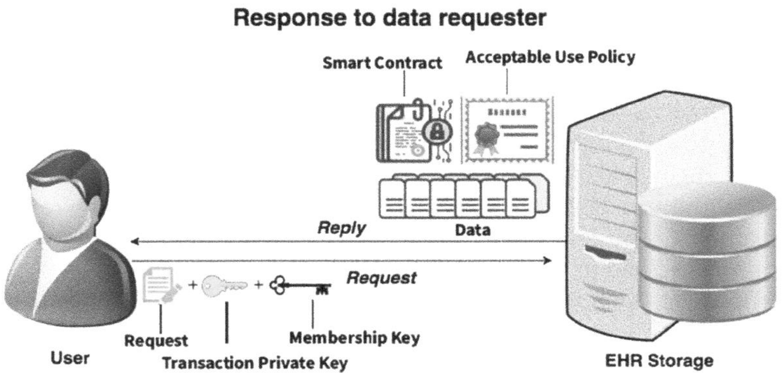

Response to data requester

FIGURE 7.2
High-level view of request and reply in our health information exchange.

or the computation period expires, destroys the data in accordance with policy. We proposed to continue our research into the subject of health data sharing by using our results as the foundation for the construction of a blockchain-based digital health platform with integrated analytics and guaranteed privacy.

References

[1] A. J. Burke. Health Information Exchange (HIE) technology infrastructure for Privacy Assurance Trustmark (PAT) test and development. In SoutheastCon 2015 (pp. 1–2). IEEE, 2015, April.

[2] Nakamoto, "Bitcoin: A Peer-to-Peer Electronic Cash System," p. 9, 2008. www.bitcoin.org.

[3] K. Abouelmehdi, A. Beni-Hessane, and H. Khaloufi, "Big healthcare data: Preserving security and privacy," J. Big Data, vol. 5, no. 1, pp. 1–18, 2018.

[4] M. G. Hansson, H. Lochmüller, O. Riess, F. Schaefer, M. Orth, Y. Rubinstein, C. Molster, H. Dawkins, D. Taruscio, M. Posada, and S. Woods, "The risk of re-identification versus the need to identify individuals in rare disease research," Eur. J. Hum. Genet., vol. 24, no. 11, pp. 1553–1558, 2016.

[5] C. Culnane, B. I. P. Rubinstein, and V. Teague, "Health data in an open world," arXiv preprint arXiv:1712.05627, 2017.

[6] E. B. Sifah, Q. Xia, K. O. B. O. Agyekum, S. Amofa, J. Gao, R. Chen, ... and M. Guizani, "Chain-based big data access control infrastructure," J. Supercomput., vol. 74, no. 10, pp. 4945–4964, 2018.

[7] C. Stagnaro, "White Paper : Innovative Blockchain Uses in Health Care," pp. 1–13, 2017.

[8] C. Mcfarlane, M. Beer, J. Brown, and N. Prendergast, "Patientory – Whitepaper," no. May, pp. 1–19, 2017.

[9] X. Yue, H. Wang, D. Jin, M. Li, and W. Jiang, "Healthcare data gateways: found healthcare intelligence on blockchain with novel privacy risk control," J. Med. Syst., vol. 40, no. 10, p. 218, 2016.

[10] Q. Xia, E. B. Sifah, A. Smahi, S. Amofa, and X. Zhang, "BBDS : Blockchain-Based Data Sharing for Electronic Medical Records in Cloud Environments," MDPI, p. 16, 2017.

[11] Q. Xia, E. B. Sifah, K. O. Asamoah, J. Gao, X. Du, and M. Guizani, "MeDShare: Trust-less medical data sharing among cloud service providers via blockchain," IEEE Access, vol. 5, pp. 14757–14767, 2017.

[12] A. Ekblaw, A. Azaria, J. D. Halamka, A. Lippman, I. Original, and T. Vieira, "A case study for blockchain in healthcare: 'MedRec' prototype for electronic health records and medical research data MedRec: Using blockchain for medical data access and permission management," pp. 1–13, 2016.

[13] G. G. Dagher, J. Mohler, M. Milojkovic, and P. B. Marella, "Ancile: privacy-preserving framework for access control and interoperability of electronic health records using blockchain technology," Sustainable Cities Soc., vol. 39, pp. 283–297, 2018.

[14] J. J. Yang, J. Q. Li, and Y. Niu, "A hybrid solution for privacy preserving medical data sharing in the cloud environment," Futur. Gener. Comput. Syst., vol. 43–44, pp. 74–86, 2015.

[15] K. Peterson, R. Deeduvanu, P. Kanjamala, and K. Boles, "A blockchain-based approach to health information exchange networks," Work Blockchain, no. 1, pp. 1–10, 2016. https://www.healthit.gov/sites/default/files/12-55-blockchain-based-approach-final.pdf [Accessed: 2019-09-05].

[16] Y. Yang and M. Ma, "Conjunctive keyword search with designated tester and timing enabled proxy re-encryption function for e-health clouds," IEEE Transactions on Information Forensics and Security, vol. 11, no. 4, pp. 746–759, 2016.

[17] J. Zhang, N. Xue, and X. Huang, "A secure system for pervasive social network-based healthcare," IEEE Access, vol. 4, no. 99, pp. 9239–9250, 2016.

[18] M. Soni and D. K. Singh. Blockchain implementation for privacy preserving and securing the healthcare data. In 2021 10th IEEE International Conference on Communication Systems and Network Technologies (CSNT) (pp. 729–734). IEEE, 2021, June.

[19] A. Zhang and X. Lin, "Towards secure and privacy-preserving data sharing in e-Health systems via consortium blockchain," J. Med. Syst., vol. 42, no. 8, pp. 1–18, 2018.

[20] T. H. Kim, G. Kumar, R. Saha, M. K. Rai, W. J. Buchanan, R. Thomas, and M. Alazab, "A privacy-preserving distributed ledger framework for global human resource record management: the blockchain aspect," IEEE Access, vol. 8, pp. 96455–96467, 2020.

[21] M. H. Onik, M. H. Miraz, and C. S. Kim. A Recruitment and Human Resource Management Technique Using Blockchain Technology for Industry 4.0, 2018.

[22] N. N. Pokrovskaia, V. A. Spivak, and S. O. Snisarenko. Developing global qualification-competencies ledger on blockchain platform. In 2018 XVII Russian Scientific and Practical Conference on Planning and Teaching Engineering Staff for the Industrial and Economic Complex of the Region (PTES) (pp. 209–212). IEEE, 2018, November.

[23] C. Dannen. Introducing Ethereum and solidity (Vol. 1, pp. 159–160). Berkeley: Apress, 2017.

8

Smart City Contracting

8.1 Introduction

Several entities come together in the workplace on a daily basis to fulfill the organization's aim. Many workflows involve the rules, protocols, and agreement set, which creates a set of operations. Each organization is structured in a systematic model to ensure that the workflow of activities is properly managed. Because of the complexity of the entities within a corporation, all working participants must be validated and activities must be monitored. Such regulations must be controlled and implemented in a transparent manner. As a result, several contracts for sub-units within corporations are made.

For these contracts, the contract execution involves various processes such as drafting, negotiation, and signing [1]. These steps are lengthy and time-consuming. Individuals and companies need to draft this information in a long legal or written form. Further consensus is reached as parties negotiate the clauses within each contract, which undergo several modifications until the contract is signed. This makes it more likely that contracts will be written in such a way that they may be easily amended to meet the demands of only one party.

Over time, the contracts that are made become clumsy as well. Searching for archived paper can be extremely time-consuming; hence, the development of digital contracts is done.

Digital contracts have provided a means by which contracts can be encoded in a usable electronic format to manage the major operations of an organization. The information created is also stored in a single storage point easily accessed by all relevant parties. To prevent people from modifying the information in a digital contract, each party participating signs the documents with their cryptographic keys to approve a particular contract.

However, this creates many problems as a single point of failure is created, making it easy for this information to be easily compromised. Although automated, the management of these digital contracts has also become a great challenge and requires a lot of effort over time. Keeping track of many contracts that are intricately related becomes nearly impossible and

DOI: 10.1201/9781003289418-8

prohibitively expensive. For example, different contracts must be inter-acted with and set up between parties in organizational operations such as recruitment and project organizations to ensure fairness and proper adher-ence to the procedure.

A single project completion may require parties to agree to meet specific demands over a written contract when completing projects involving a diverse variety of people with different jobs and skill sets. Contracts must be managed, controlled, and interacted with properly since they are important in giving guidelines that regulate how main operations are carried out in each firm.

There is also the need for units such as the human resource units [2] to have better control of the contracts they manage since they are central to an organization and take care of complex processes, such as recruitment, assigning responsibilities, and arranging compensation plans. They are also responsible for ensuring employee and employer safety [3, 4], wages and salaries, contract creation, and regulating relationships with exter-nal parties such as trade unions and outsourced contracts. Contracts for these tasks are written down and set up in multiple interlinked contracts to serve various purposes. In the case of disagreements [5], it becomes a more complicated issue.

Currently, third-party means of dispute resolutions are also embedded in these digital contracts to provide a means by which disputes between contracting parties can be resolved. Also, contract drafting still remains a challenge due to the number of considerations to be taken in designing an effective contract. Even if this is done, before making these contracts digitized, they still remain unenforceable since many parties must coop-erate before the agreement and rules set in place can before the contract can be activated [6]. This thus makes the contracts designed unmanage-able, difficult to enforce, and prone to several alterations, modifications, and attacks.

With the emergence of the blockchain, there seems to be an option in the way contracts are done with the introduction of smart contracts that pro-vides its inherent contract verification mechanism and is Turing complete and designed with better enhanced automated features. With its decentral-ized nature on the blockchain, it becomes harder for contracts deployed to be changed and altered as easily as traditional digital contracts. However, the majority of work done for blockchain-enabled companies still employs traditional digital design approaches that rely on trusted third parties for enforcement. In addition, few works have been done to develop a well-adaptive framework for organizations that are compatible with smart con-tracts' advantages. In the case of managing contracts, many works have been unable to utilize the ability for smart contracts to interlink [7] with each other securely to provide an effective mechanism for managing multi-contracts among multiple parties within multiple organizations. The dispute resolution among employees and employers within an organization suffers from the lack of trust among contracts developed using the current system. Various parties are left cheated in several instances due to parties' inability to participate in the contracts developed and contracts written.

We propose a solution to these issues by utilizing the capabilities of blockchain-based smart contracts [8, 9] to create a viable framework for various multi-contracting parties within a company. We specify novel rules and constraints that blockchain-based smart contracts must follow to meet contracting parties' needs. We develop a new approach to utilizing smart contracts within single and multiple organizations by creating a mechanism within each organization's workflow, such as organization registration, recruitment, employer and employee negotiation, and dispute management [10].

Through the use of decentralized oracles with the blockchain network, a novel recruitment and application system is developed to provide a trustless environment where trust is established between employer and applicant and participant within the application process. Trust is also ensured over the negotiation system between employer and employee and outsourced workers by leveraging the features of the blockchain contracts in a complex multi-contracting system [11, 12] that is self-verifying and provides security, transparency, and trust in several contracting parties.

1. We provide a novel approach by looking at pertinent workflows of organizations, for instance, recruitment, outsourcing, and arbitration [13].

2. We develop a mechanism for parties to interact over the blockchain network in a trustless multi-contract setup.

3. We offer a contract management system that automatically organizes and interlinks complex contracts [14] in a way that allows for easy access and coordination among different parties.

4. We create a decentralized blockchain-based oracle network with a transparent recruitment and employment [15] process that is based on the blockchain and provides a fair way of hiring parties.

5. We create a new blockchain-based dispute resolution [16] mechanism to ensure that disagreements between contracting parties are resolved quickly, fairly, and transparently without the involvement of a third party.

6. We develop a hierarchical, interlinking scheme for which contracts are arranged and can track and accessed easily.

7. We present a novel method for achieving multi-organizational collaboration across various contracts. We accomplish this by utilizing smart contracts' interactive capabilities.

8. We provide a way for employees, employers, and labor unions to interact with one another through contracts created on our proposed blockchain-based system.

9. We consider contracting in ex-facto situations. We make consideration for these situations and their occurrence over the blockchain in arbitration procedures.

10. We also present mathematical arguments to examine the consequences of moving decision-making power [17] and leverage among numerous contracting parties on the decision taken.

8.1.1 Incentive Theory

In our proposed proposal, we apply incentive theory to better understand the behavior of contracting parties in an organization, as each party seeks to increase their reward during the negotiating process. It also becomes useful in [18] our efforts to develop a set of norms that will benefit all parties. The basis of incentive theory is a process in which a principal party P devises an incentive scheme to persuade the informed party, the agent, to reveal information (adverse-selection model).

The strategy is based on two key assumptions: the principle is underinformed to some extent, and the concealed variable X is unknown to him. They also have no idea about the variable's probability distribution or the agent's preference structure [19]. As a result, the principle can put himself in the shoes of the agent, anticipating the latter's reactions to the range of possible remuneration schemes and choosing the one he or she wants among those that are acceptable to the agent. This capability is provided by the blockchain's proof of work consensus method.

In this scheme, the contracting parties enact a transaction [17]. To verify the transaction, the contracting parties cryptographically solve a given puzzle to prove the transaction made. An honest consensus scheme is created since there is a low or no probability for two nodes to solve the cryptographic puzzle with a high difficulty setting. The parties do neither know the cryptographic puzzle nor the next person to solve among an increased number of nodes on the blockchain network. The information verified is stored on the blockchain due to the trust established among the nodes. The blockchain provides an incentive mechanism to the parties that solve this puzzle known as miners on the blockchain. A fair mechanism is created for all contracting parties who have to play honestly to increase their payoff.

8.1.2 Transaction Cost Theory

We make use of transaction cost theory in this work to understand the outcomes of an incomplete contract [20, 21] This describes a set of contracts that, although satisfying the present requirements, cannot account for future unpredictable scenarios. Thus, in ex-post-facto situations, contracts defined may have difficulty covering issues that may arise. It is thus used in the area of non-savage rationality. This means that the agents exist in a world of limited liabilities in which they cannot calculate and do not know the structure and set of problems that may arise. These agents are confronted with "radical" uncertainty δ, which prevents them from composing a complete contract.

Agents, on the other hand, must include measures to command the actions of each ex-post in order to ensure coordination despite the incompleteness of their contracts. The contract allocates decision rights to one agent to influence the actions of the other.

- Single party *A*
- Multi-party *A, B,..., n*
- External Party *C*

Because of the blockchain's immutability, we use it in our plan. A contract cannot be amended or altered once it has been distributed on the blockchain network. As such incomplete contracts being drafted on the blockchain must utilize certain pre-conditions to ensure that contracts drafted satisfy expost scenarios and can be negotiated upon after situations external to their deployment changes.

8.1.3 Contracting

As we have seen in our research, most contracting decisions are based on a standard discrete choice problem. Parties to a transaction will opt to contract if the expected gains are higher than the expected gains from structuring the transaction in another way. This is taken care of by the blockchain's consensus protocol and the fair environment [22, 23] that it creates for numerous parties to contract. We formalize our concept as follows:

$$G^* = G^c, \text{if } V^c > V^a, \text{and} = G^a \text{ if } V^C < V^a \tag{8.1}$$

G^c is the primary contractual value. G^a is the alternative contracting value, V^c and V^a are the transactor's presumptions in each contract generation, as well as the associated values that make up the transaction under the leading and alternative contracting values. G^* is the contracting value that was picked as the best.

It is necessary to add to the preceding arguments,

$$V^c = V^c(X, e_c) \tag{8.2}$$

$$V^a = a^X + e_a \tag{8.3}$$

X is a vector of observable properties that affect contracting gains under predetermined arrangements e_a and e_c. These are mistake phrases that could indicate contracting parties' misunderstandings concerning the verifiability of the presumed values, V^c and V^a. We presume that the preceding relationships may be expressed linearly as the likelihood that contracting will be selected over all other contracting options.

8.1.4 Contract Duration

The issue of contract duration is another contracting decision to consider on the blockchain. Multiple parties can also be seen as deciding how so many periods their contract should span (if any). In the absence of a contract duration, a limiting case of zero contract duration will be created. On the other hand, the contract length could be represented as a collection of discrete options in which several parties transacting select whether to participate in a contracting process for every future period. T the potential could be used to express this discrete choice decision (possibility infinite).

This is how we say it:

$$\max_{\tau} V^c(\tau) + V^a(T - \tau)$$

This denotes the length of the contract T, the length of time that the parties' connection could last. The cumulative value of contractual exchanges made over the contract's duration is referred to as $V^c(\tau)$. The $V^a(T - \tau)$ represents the value of activities taken in the period following the contract's expiration. The one number that satisfies the equation is referred to as the optimal contract τ^*.

$$V_\tau^c\left(\tau^*\right) = V_\tau^a\left(\tau^*\right)$$

The participants can keep increasing the contract term till the value of transacting under a contract for an additional period is just equal to the value of transacting without the need for a contract during that time. Of course, this is reached over a long period. However, as the parties begin to contract over the blockchain over a long time, this is ensured [24]. The issue of fairness and security is also considered for each contract given in each time period.

8.1.5 Decision Trees

To offer a hierarchical structure for contracts, we use decision trees. This promotes clarity when making and examining decisions, as well as monitoring their consequences [25, 26]. Assume that there are n decision nodes set up across the blockchain network among numerous contracting parties, and that each decision node has an action set A_i describing the choices that can be made at that point over the network. The set of pure strategies is formed when some or all of the sets A_i are identical. The cross product of all the action sets yields S: $S = A_1 A_2 ... A_n$. A strategy with no randomization is known as a pure strategy.

If there are about three decision nodes for a series of actions, an action set is logged on the blockchain at each time t: $A = a_1, a_2$, $B = b_1, b_2$ and $C = c_1, c_2$.

In this scenario, we express the set of pure strategies as

$$S = a_1 b_1 c_1, a_1 b_1 c_2, a_1 b_2 c_1, a_1 b_2 c_2, a_2 b_1 c_1, a_2 b_1 c_2, a_2 b_2 c_1, a_2 b_2 c_2$$

This serves as a sequence of decisions made over the blockchain network. The process, though, may come to an end before all possible decisions have been made. The current gratification is foregone in the hope of a future benefit. This is seen as strategic behavior. Secondly, the behavior of the other party is taken into account. In our scenario, all participants have the capacity to modify their behavior at any time. We were representing a game similar to that shown in Figure 8.1. This gives us a decision tree.

Contracting parties' decisions on the blockchain are shown as little filled circles. Each action at a decision node has its own branch leading away from it. When all decisions are made, one reaches the tree's end. The payoff for taking that path is documented and published transparently on the blockchain.

Mathematically, each decision node is marked by an indicator set $I = 1, 2, 3, ..., n$. At node i, the number of available action set k, which is different at each decision point, is created. The probability vector determines

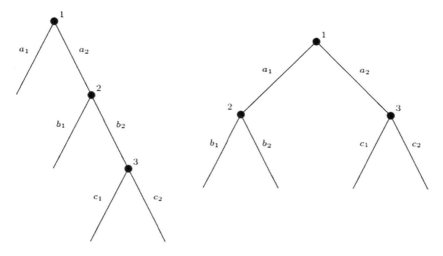

FIGURE 8.1
Diagram a decision node tree structure that provides a set of actions and outcome.

everyone's behavior at node *i*. p_i where $p_i = \big(p(a_i,1),p(a_i,2),\ldots,p(a_i,k_i)\big)$ and $p(a_i j)$. Action a_i, $j \varepsilon A_i$ is chosen based on this probability, that is, if decision node *i* is reached over a period of time *t*, a collection of probability vectors β is formed.

$$\beta = p_1, p_2, \ldots, p_n$$

We stick to a game of pure strategies at first. However, due to the variable set of actions on the blockchain, there are moments where players approach the game with a set of mixed strategies in uncertain situations, which is created over time *t*. This continues for the set of contracts used in each game. Therefore, randomization of actions occurs due to the mixed strategies, bringing about a set of behavioral strategies that occur as the decision tree is traversed.

In this instance, due to the nature of the blockchain, providing transparency for all party members, we observe a sequential set of decisions made by contracting parties.

Theorem 1
Every mixed strategy has a behavioral description, and every behavioral strategy has a mixed description.

Proof
When interaction with a contract $s \varepsilon S$, a person adopting a pure approach will traverse through a set of decision nodes $I(s) \varepsilon I$, and take an action $a^i(s) \varepsilon A_i$ for each $i \varepsilon I(s)$. A particular behavioral strategy β would dictate choosing that action with a degree of probability $P\big(a(i)(s)\big)$ at each point of the decision node $\varepsilon \varepsilon I$.

Then, β is a mixed strategy representation of

$$\sigma_\beta = \sum_{s \varepsilon S} p(s)s$$

Provided $p_\sigma(i) = 0$, the likelihood of selecting a^i at *i* can be expressed as

$$p\big(a^i\big) = \frac{p_\sigma\big(a^i,i\big)}{p_\sigma(i)}$$

Any set of probabilities $p(a_i)$ taken with the game if $p_\sigma(i) = 0$ for some decision node *i* being played among the set of contracting parties with $a_i \varepsilon A_i p(a_i) = 1$ will suffice.

8.2 Contract Setup for Multiple Organizations

We created a collection of smart contracts that define the features that allow enterprises to communicate via the blockchain. We created such contracts to build a blockchain ecosystem in which corporations may collaborate in a fair manner. We start by establishing a contract to track the registration of all elements and units that will interact. This is referred to as the *RegistrarContract*.

This contract is in charge of launching other contracts and taking note of all contracts that have been established and launched by assigning an id to them and preserving their blockchain addresses. It also takes care of the registration of organizations, human resource units, employers, employees, trade unions, outsourced contractors, and subcontractors.

We also create a contract for the entities in charge of the day-to-day operations of the organization and the specific function they are to follow within the ecosystem. We define them as *MainContracts*. These include *OrganizationContract*, *HumanResourceContract, EmployerContract, EmployeeContract, TradeUnionContract, OutSourcingContract, SubOutSourcingContract, OutsourcedContractorContract*, and *SubContractorContract*.

This contract is accessed by entities performing roles assigned to them. Aside from the OrganizationContract, which represents units of organization inside our blockchain-based ecosystem, these contracts build an access control structure to regulate their functions within an organization.

The rest of the MainContracts are in charge of recruitment, employee wages, outsourcing, contract management, and trade arbitration inside each firm. Each smart contract has a set of functional procedures that define how the many players functioning as nodes on the blockchain network will operate [27]. We define operational contracts that allow participants to work together on the blockchain.

We created more contracts to manage applications within each enterprise. These are oracle-based contracts. They connect the blockchain to the outside world and customize its use to an organization's needs.

In our work, we used oracle-based contracts to manage the recruitment process for entities within a company. The application-based contracts are as follows: *ApplicationContract, RecruitmentContract, ReputationContracts, AggregatorContract, OracleTokenReceiver, EmployerContract, OracleCLient, Oracle AggregatorRequestor, WithdrawalInterface*, and *AggregatorInterface*.

We define it as a series of contracts that deal with collaboration and conflict between internal and external parties, such as trade unions, outsourced contractors, and subcontractors. ContractRankingContract, NegotiationContract, DisputeResolutionContracts, and ArbitrationContract are just a few of the contracts available [28].

8.2.1 Registration

The RegistrarContract is used by an organization applicant OA_i to make a blockchain application. All organizations on the blockchain are registered as $org[O_i,...,O_n]$ by the RegistrarContract. The RegistrarContract opens a session as OrganizationRequestSession ORS and sets a unique ID as ORSID once the application is set. After the ORS is set up, the requirement Req is created, which is a set of needs depending on the organization type. The RegistrarContract submits the RegistrarContract requirement.

Each blockchain-based corporation must adhere to the requirements and set of criteria outlined in the smart contract installed on the blockchain. The data submitted is validated against the standards defined once it is supplied to the RegistrarContract. An instance of the OrganizationContract is created on the blockchain after the criterion is met. The remaining two MainContracts are created. EmployerContract and HumanRessourceContract are the two contracts in question. This is given a secret key and a public key for the EmployerContract. After that, his status is changed from applicant to employer.

He also has the secret key to vote on the assignment and approval of the HumanResourceContract to many entities, as well as the request for the assignment of that position by other actors, as an employer. Within each contract entrusted to them, all parties have access to the numerous modular functionalities. This gives the organization functional control as well as good access control.

8.2.2 Recruitment

The ApplicationContract receives a request from an applicant. He also pays a little charge to the contracts department. However, this may not be necessary to make the full functionality of our proposed idea. This is done to avoid a denial-of-service attack and prevent an overly submission of required data Reg_{data} submitted. This fee is also used by the ApplicationContract to pay for the Oracle Nodes service. The contracts participating and interacting with the oracle nodes include the *ReputationContracts*, *AggregatorContract*, *OracleTokenReceiver*, *EmployerContract*, *OracleCLient*, *OracleAggregatorRequestor*, *WithdrawalInterface*, and *AggregatorInterface* contracts.

An organization O_i establishes a hiring procedure for new workers Emp_i to Emp_n. On the blockchain network, an implementation of the RecruitmentContract is launched. This contract outlines the standards and procedures that each organization should follow. This is between the organization and the applicants in the recruitment process. Each applicant must adhere to the RecruitmentContract's established set of agreements. The HumanResourceContract connects and monitors the recruitment process. A candidate App_i requests to participate in the application procedure.

The applicant App_i submits a set of requests $App\,Re\,g_i$. All required information from candidates is supplied as $Re\,g_{data}$ and given to the *ApplicationContraact*. This contains his credentials, skill set, and the position he wants to apply for within the company O_i. This must correspond to the set of conditions Req that each *RecruitmentContract* applicant App_i must meet. These requirements involve a set of highly specialized skills, attributes, competencies, and functions that the applicant must meet. This also specifies the responsibilities of the applicant once recruited. To store the applicant's competency information, a system for saving global qualification or competency information for each application is built.

The blockchain's decentralized nature serves as an efficient place for storing global qualification or competency information for each applicant, providing transparency from which analysis can be made. When the data is received on the blockchain, an application session $App_{Session}$ with a unique id $App_{Session}$ is started. The data is sent to AggregatorContract, a blockchain-based aggregator with a collection of oracle nodes that pulls data from a variety of sources. The data gathered is used to create a list of applicants' global qualification ledgers.

The AggregatorContract, which is built on the blockchain, uses API entry points to request data from data sources. Queries are then issued to a number of nodes n on the data provider's node. Across numerous data points, each AggregatorContract looks for replies from each node it connects with. The ResponseChecker within the AggregatorContract looks for responses from each node and collates this information, looking for metrics relevant to its study. The data is then checked for accuracy to eliminate any fake data feeds.

The ReputationContract also contains a ReputationMetricsIndex, which acts as a check to prevent false and erroneous data. When the data collected meets the ReputationMetricsIndex, it is logged onto the blockchain's Global Qualification System GOS for evaluation by employer E_i. If the applicant is not accepted, his $Application_{Status}$ on the blockchain is converted to REJECTED. On the blockchain, a notification is transmitted to the HumanResourceContract. An instance of the EmployeeContract is deployed on the blockchain once the notification from the EmployerContract is received. The applicant App_i is assigned an ID, and his status is updated to CONFIRMED. He is then added to the $Emp_i, \ldots, Emp_n$ group of employees within the O_i organization.

The EmployeeContract specifies all information concerning payment, obligations, consequences, and negotiating procedures. A receipt is produced if the transaction is accepted, and the transaction information and details are logged and disseminated to all nodes within the organization. This serves as a reference point to which each employee will be held accountable, as well as each activity taken within the firm. The smart contract is implemented on the blockchain, and it is this that allows the two parties to interact.

In ensuring that all parties effectively make decisions, our blockchain-based approach ensures that employees are systematically provided with the needed information on matters that concern them through the transparency of the blockchain. Details of economic employees' involvement and activities are recorded on the blockchain. For instance, employees' involvement in initiatives is recorded as an encouraging start on the blockchain. All participatory entities on the blockchain see this. GoodAssistant and other indicators are based on excellent conduct, which is also recorded on the blockchain. This allows you to keep track of what's going on in each organization. We also observe whatever is going on in each organization in this way. A high level of employee participation and employee governance is effectively enabled in this way on the blockchain.

By this means, the blockchain provides a means to deploy man power effectively and involves the employees in the organization's daily practice and activities. The reward and penalty mechanism set up in the initial recruitment process also ensures that efforts are incentivized and monitored on the blockchain.

8.2.3 Outsource

Organizations must outsource services beyond their jurisdiction in order to assure accuracy and productivity. This is the everyday practice where organizations collaborate remotely from different locations to achieve a particular project. This means different organizations must come together. Multiple contracts are set up for this. Various interests are satisfied, and several negotiations are made. Delivery times and roles and responsibilities must also be developed for multiple organizations or entities within various organizations. This creates a complex system where multi-contracting becomes very difficult.

The outsourcing company requests an Outsourcing Service OTS with an Outsourcing ID of OTS_{ID} in our scheme. A new OTS with a unique OTS_{ID} must be formed for each on-chain outsourcing. An OTS's participating organizations are recorded in the blockchain. Because the HumanResourceContract is so important to the organization's operations, both sides must be able to trust one another. We create an outsourcing system that ensures complete transparency for all parties involved. After the outsourced company O_a has set up the OTS, *the OutsourcingContract* sends a request Req to the required organization.

The required organization is designated as O_b. When O_b receives the request Reqst, he or she must approve the request. The scope-of-work SC is included in the request. The SC defines the name of the organization, its details, the entity required, the skill set required, timelines, and the blockchain addresses of the entities required. When O_b receives the request, he authorizes it. The outsourcingContract enlists all outsourced contractors for the project under the OTS_{ID} once the Reqst is granted.

This combines the information from each outsourced contractor into the *OutsourcingContract* with a specific, OTS_{ID}. OutsourcingContract also collects the necessary information or profile of the Outsourced Contractors *OCP*. After that, an instance of *OutsourcedContractorContract* is created for each of the parties. This works as a contract via which they can interact with one another. There are additional rules for each of these outsourced contractors in the contract.

A time frame t is defined for the task indicated in the SC once the OutsourcingContract is achieved. The OutsourcingContract sets the work status WS as complete if the requirement Req established for the task in SC and the time frame t are met. If the contested contract fails to meet the condition, DisputeContract is launched. Rewards and penalties are enacted for parties on the blockchain once the issue is resolved.

A mechanism for rewards and penalties is built up in each contract based on the time frame set. We analyze cases in which numerous companies are required to complete the task. First, when many organizations $OrgO_i, \ldots, O_n$ outsource, each one will require a set of OTS to be developed and stored as Outsourcing Service Tracker OTST. This assigns the number OTS_i to all OTS on the blockchain. The OutsourcingContract is in charge of managing and controlling this. As a result, a complicated structure, including several organizations performing a single activity within a very complex eco-system made up of multiple partners with multiple contracts is developed. The blockchain takes care of this in a secure and efficient manner. Without the involvement of third parties, enforceability and verifiability are immediately realized among participating parties.

In this way, a complex structure is created involving multiple organizations performing a single task within a highly complex eco-system that is made up of multiple parties with multiple contracts. This is managed securely and effectively by the blockchain. Enforceability and verifiability are automatically achieved among participating parties without third parties.

8.2.4 Subcontracting

For any assignment delivered or outsourced from O_a to O_b, a portion of the task may be sent to a third-party entity unknown to O_a in areas that may be beyond the specialization of O_b. For instance, many organizations outsource their work to other subcontractors to deliver on a single project in construction work. These outsourced responsibilities are frequently beyond the competence of the firm to which the contract was assigned.

Nevertheless, many contract-dependent decisions, such as compensation and job timeliness, are frequently reliant on several parties' consent and fulfillment of these conditions. Our plan combines these features into our proposed solution. For each *OTS* generated on the blockchain, we establish an interconnected structure of subcontracts. The Merkle root hash of the blockchain offers a way to track back transactions created, establishing a hierarchical system, thanks to its connected node system.

We established a service known as a Suboutsourcing Service *SCS* to monitor the *SubContractors* that Outsourced Contractor will contract for a particular OTS_{ID}. Since this information is emitted on the blockchain, all parties involved with a particular project with the specified OTS_{ID} can view the details involved and track the relevant details once the task is outsourced to a subcontracting organization O_c from an organization O_b.

A similar setup as the setup for the outsourced contractor is initiated. The scope-of-work *SC* for the particular *SCS* identifier SCS_{ID} is submitted to O_c by the *SubOutsourcingContract*. Once received and approved, all subcontractors' details are enlisted and are submitted under a specified SCS_{ID}. This connects all the enlisted subcontractors' details stored as *DC* on the blockchain for the given Subcontractor Service.

For all enlisted subcontractors, an instance of the SubcontractorContract is started. The NegotiationContract creates a negotiation process when it is instantiated. The Subcontractor Service is closed if the negotiation cannot be concluded. The time period given in SC for the specified SCS_{ID} is instantiated once the negotiation is completed. The requirement Req defined inside the project SCS_{ID} is checked once the time frame within SC is reached. There is a reward transfer if the subcontractor meets the criterion.

If he is unable to meet the condition, the DisputeContract contract provides for the initiation of a dispute process. The details of all transactions are added to the DC once the dispute resolution procedure is completed. This is added to the SubOutsourcingContract's scope-of-work SC and submitted. The SCS_{ID} of the given service is subsequently recorded as SCS_{ID} in the blockchain's Subcontractor Service Tracker.

8.2.5 Negotiation

Due to various demands within an organization, there is the need to create an environment where needs and claims are met, and parties can negotiate. A collective bargaining scheme is created on blockchain to assist all parties. This agreement includes pay and condition of employment. This helps them make effective decisions. This provides information as well as knowledge of the economic factors that affect the performance of the organization. The blockchain provides a mechanism of certainty and fairness, ensuring that such requests are formalized through the smart contract provided. The NegotiationContract is in charge of the negotiation process; any request transmitted to the blockchain by a party is referred to as request *R*. Employees, outsourced contractors [8], and subcontractors on the blockchain are then received by the employer or negotiating party and alerted on the blockchain based on the request.

This creates a valid way of verification for fairness and openness. The NegotiationContract designed for each service also gives the reply and terms

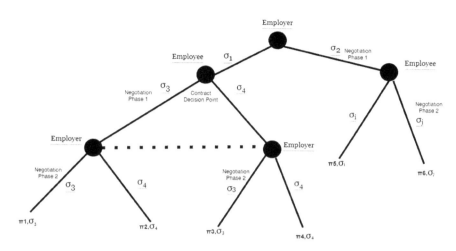

FIGURE 8.2
Diagram showing decision node of each negotiation between employer and employee.

for acceptance and rejections. This is visible to all on the blockchain as seen in Figure 8.2.

The collective agreements are legally enforceable with the use of smart contracts since every breach by the employer, employee, and contractors and subcontractors are immediately punishable by the *NegotiationContract* and *DisputeContracts* set. Numerous blockchain sessions are started until a final decision is reached on the blockchain. The contract's procedure calls are then started to continue the flow of service.

Since the contracts are designed as smart contracts, the smart contract formalizes these two parties' collaborative agreement. As a result, whatever arrangement the employer and employee reach, once written on the smart contract, becomes immediately enforceable, thus creating ease in enforcing agreements. Our blockchain scheme also provides another means of negotiation, even in informal negotiation circumstances. For instance, in situations where an employee decides to sue his employer in reducing pay, which is sometimes possible in certain scenarios, a lack of evidence to support the claim and the agreement between the employer and the employee may usually be seen as an informal contract agreement.

This is unusable in courts and therefore unenforceable. However, a collective agreement on the blockchain makes it easy for them to engage in an informal agreement outside of the court that the blockchain enforces. Therefore, it is easier for all parties to come together on the blockchain in complex collective agreements and engage in an enforceable agreement structure on the blockchain. This takes away the oral agreement's existence and replaces it with something more enforceable and implementable on the blockchain. Since all members such as union and employers form part of

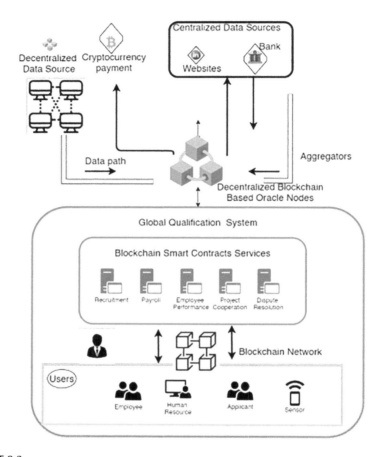

FIGURE 8.3
Diagram showing a single unit contracting organizational structure interacting on the blockchain and external environment.

the agreement and all actions on the blockchain are viewable, transparency provides collective participation as seen in Figure 8.3.

Therefore, a breach of contract is evident. In a situation such as unfair dismissal, parties over the network can pursue the enforcement of the contract and thus ensure that remedies are given to the dismissed employee but initiating enforcement procedure enabled by the blockchain.

8.2.6 Managing Trading Unions

An organization to which an employee belongs certifies the recognition of each union on the blockchain. Both the *HumanResourceContract* and the *EmployerContract* must certify each trading union's recognition. The

HumanResourceContract HRContract manages information and logs it on the blockchain. Another contract is established among the employee, the employer, and the organization, known as the *TradeUnionContract*. The contract arrangement between the employer and the trading unions is created to secure employees' rights, and this information is then logged into the blockchain. The transparency of membership between the organizations and employee prevents any poaching from occurring.

For instance, in the case where it is recognized that unions have organized activities at places where another union's members, it is realized that upon verification, penalties are enforced into the blockchain-to-blockchain addresses enlisted in the *TradeUnionContract*. As a result, regulating the actions of participants of a certain union in the event of a contract breach is automated. Each employee of a certain organization O_i must first register with a specific TradeUnion before the collaboration can begin.

The employer and the trade union, respectively, approve the application through the EmployerContract and the TradeUnionContract. Once this is accomplished, a blockchain receipt is generated. Each trading union is not only confined to a single employer but also includes a list of additional employers with contractual personnel. The TUC (Trading Union Contract) lists all of these groups. For each trading union registered on the blockchain, an instance of TradeUnionContract is produced. There is a link made between the TradeUnionContract, EmployeeContract, EmployerContract, and OrganizationContract after the TradeUnionContract is set up. The smart contract establishes an automated link between these entities that is both possible and beneficial to all stakeholders.

We refer each organization to a specific trading union with the id TUC_{ID} because the OrganizationContract itself provides the details of the organization id, the name of the employer, roles, employee ids, and other information of the company. We do this by linking the TradingUnionContract and OrganizationContract that an organization O_i has access to. This registration and linking aid in the establishment of a formal framework for the certification of unions. As a result, the blockchain is used to certify contracts between these firms and the employer.

This helps prevent conflicts in moments of disputes. A notification mechanism is also created for both parties as they transact to provide for the employee's needs in the organization and ensure the employee's safety under their governance. Since the employee is likewise linked to each organization and trading union simultaneously, a unique way is created for the recognition of trade unions by employers. A system of negotiation of terms and conditions of employment and procedures for managing disputes is created, providing a fair environment for all parties to cooperate and work together in a more efficient blockchain-based eco-system.

8.3 Dispute Resolution on the Blockchain

On the blockchain network, decisions are made faster and at a lower cost, which substantially shortens the dispute resolution process. Due to the nature of the blockchain, every action is immediately verified, and multiple parties see transactions on the blockchain network. As such, in the dispute resolution process, where speed of decision is needed and immediate provision of evidence and trust is needed, the blockchain provides an interface for this kind of process.

The blockchain also provides a good framework for effectively managing multi-party and multi-contract arbitration. Since the blockchain can manage complex identities, submissions effectively among the employer, employees, and outsourced contractors and subcontractors, the needed proofs, submission, and pleas are securely handled. Therefore, an in-built mechanism for many participants to participate in a single arbitration process, transparency, and a faster rate of decision-making is greatly enhanced.

We prevent arbitration in conflicts by guaranteeing that all significant alternative dispute resolution procedures are addressed on the blockchain. Instead of going straight to arbitration, we develop a multi-tier conflict resolution system that avoids leapfrogging some procedural steps and avoids going straight to arbitration, which can be expensive to compute. The initiator is given the choice of entering any of the pre-arbitration modes available.

DisputeResolutionContract: Other subcontracts, such as the MediationContract and ArbitrationContract, are made available as a result of this. The blockchain automatically launches the arbitration step whenever another set of pre-arbitration processes, such as mediation, has been exhausted. A pre-arbitration functional module is created from which he or she can choose from several arbitration modes. Once the initiator selects this, a request is sent to the other participants on the blockchain to approve.

8.3.1 Mediation

We build a Mediation Smart Contract MediationContract to regulate the mediation process's key procedure during the mediation process. The MediationContract is linked to the ArbitrationContract and allows access to the arbitration phase once the contract's functional modules have been used up. During the mediation process, multiple parties submit multiple claims to a specific violation of contract that must be resolved. In many scenarios on the blockchain, not only does the employee make a claim, but the trade union also claims breach of the *TradeUnionContract* and other contracts associated with it.

The claims are made in order to defend their employee and resolve the grievances made by the employee. The *claim()* and *offer()* functional procedures inside the Mediation Contract are used by the parties to pitch their claim and offer in our scheme. These offerings are used to mediate the situation. Because collective bargaining agreements are enforceable, decisions are instantly implemented. The smart contract formalizes all parties' collective agreements. The use of centralized entities is replaced with automatically enforceable agreement structures.

The blockchain network provides for all parties acting as nodes on the blockchain network. This contract's focus is to resolve situations of breach of contracts such as unfair dismissal, unpaid wages, which can be meditated upon, and an agreed settlement made in resolution to the dispute by all parties. Until an appropriate offer is made, each party receives many claims and offers. This is then sent to the blockchain network as transactions. The contributions are cryptographically validated before being added to the transaction's blocks on the network. The results are disseminated throughout the network and can be seen by multiple nodes. Transparency is therefore achieved in the mediation process as seen in Figure 8.4. A consensus among multiple parties is reached.

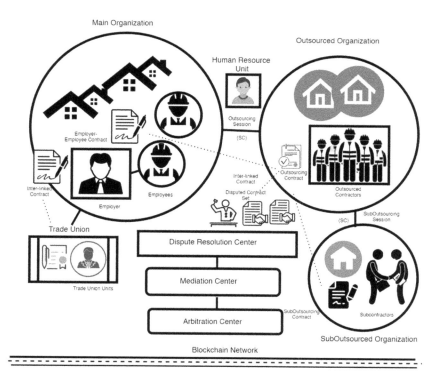

FIGURE 8.4

Diagram showing a functional modular structure for our proposed inter-contracting blockchain-based multi-organizational eco-system.

Nevertheless, there may be circumstances in which parties acting as nodes on the blockchain network wish to enter the arbitration phase without first resolving any disputes. In this case, each party submits a token to be recorded on the blockchain. One side desires that the activation phase be started right away. This is only possible if all parties agree to pay to enter the arbitration phase and reach an agreement among all parties involved in the dispute.

8.3.2 Multi-Arbitration and Multi-Contracting Arbitration

Multi-party arbitration involves disputes between multiple entities [25, 26], for instance, between an employer and members of an organization. This also includes members of the organization and themselves. Likewise, multi-contracts refer to disputes surrounding various contracts written between parties within an organization and multiple organizations. In some organizations, contracts created can be bilateral contracts among employers, employees, outsourced contractors, and subcontractors.

In some circumstances, disputes may arise when there is a conflict between the employer and the employee and outsourced contractors and subcontractors. For instance, work can be outsourced to other parties in a design organization, although all tasks' intended combination is to achieve a single goal in the organization. A more complex dispute scheme arises when the outsourced contractor further sues the secondary contractor to whom he relayed the work. For instance, the employer may sue the outsourced contractor for delayed damages resulting from late completion and delay work. Also, conflict may arise between the employee and the employer from delay in payment and payment conditions. These may likewise result from delay damages caused by one party's inability to do his or her work well. Thus, this long strain of disputes requires a dispute management mechanism, something we leverage the blockchain and its inherent smart contracts to utilize as shown in Figure 8.5.

8.3.3 Arbitration Scheme

We create the Arbitration Contract *ArbitrationContract*, a contract that includes an arbitration plan for all parties. The arbitration system deals with single contracts, also known as bilateral contracts and multi-contracts, because the procedure functions are established in each contract. A multi-contract can involve a contract between an employer and employees from different companies on the other side, such as a contract between an employer and other outsourced or freelance companies on a partial contract basis. It is understandable sometimes that the employer may prefer a single contract for a single party. However, complexity among parties in the workplaces will not make this feasible since parties and negotiations are multifaceted.

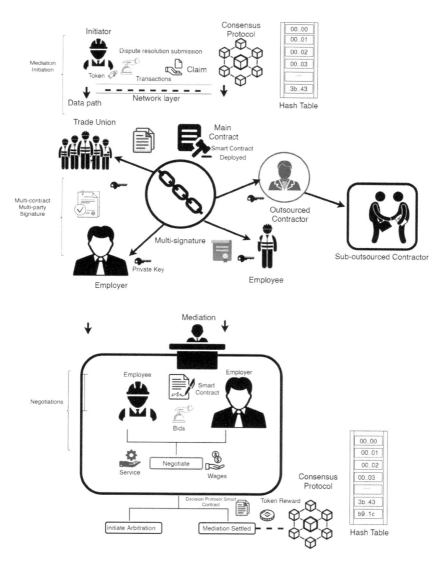

FIGURE 8.5
Diagram showing [29] data path for all processes on our proposed blockchain-based contracting framework.

There are valid reasons why the employer will not allow multiple parties to engage in his arbitration scheme. The inclusion of multi-parties into the arbitration process may delay the resolution process [30]. The cost of dispute resolution will also likewise increase. However, to account for the whole scope of dispute and provide an effective dispute mechanism, we will look at the scope of all parties coming together. Hence, our concentration

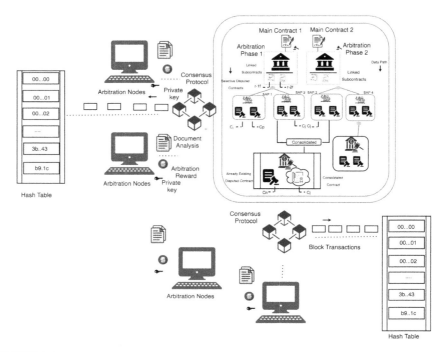

FIGURE 8.6
Diagram showing a blockchain-based arbitration process with contract management and dispute consolidation mechanism.

will look primarily at multi-contract arbitration for multi-party arbitration events as shown in Figure 8.6.

A single ArbitrationContract is used to start the arbitration procedure. This is where the arbitration process is managed. We set up a single arbitration phase in this process. An arbitrating party sends a single request Req to the blockchain during this phase. This starts the blockchain-based arbitration process. A payment r token is created and recorded on the blockchain. An arbitration phase is created by *Abphase*, which contains all of the pre-requirement R_e and procedure calls for an arbitration phase.

The blockchain addresses of parties needing to participate in this arbitration round receive an invitation call E. Each party, whether the employer, employee, or outsourced contractor, is invited based on the pre-contract requirement. After receiving this invitation, he or she must approve v to proceed to the arbitration process. If you do not, you'll have to pay a penalty of c. Any entity participating in the arbitration phase can invite an unlimited number of people as long as they are accepted under that specific arbitration id with Ab_i. After all of the participants' entries have been authorized, a time

T is established for them to present documented evidence. The arbitration process begins when T is completed, with evidence being analyzed on the blockchain.

All parties must first agree to a multi-party arbitration by signing the *ArbitrationContract* before the procedure may begin. This is accomplished using a blockchain-based multi-signature signing procedure. Although this is sufficient for simple multi-contracting and multi-party arbitration involving fewer parties, the arbitration procedure becomes considerably more complicated in multi-contract conflicts [31]. This occurs when non-identical parties, such as parties between organizations, are bound by two or more arbitration agreements. As a result, it is critical to take into account all contracts and subcontracts during the arbitration procedure. To address this problem, we develop a framework for combining arbitration and arbitration.

8.3.4 Arbitration Calls

In dealing with the contracts involved, it must be noted that the arbitration call must not be out of the scope of the designed contracts. Even if all the multiple parties are bound by a single contract with a pre-defined clause designed as a functional procedure on the blockchain, it should be assumed that in accepting the case of a multi-party contract, parties have already consented to sub-contract disputes under the specified contract. This is still the case even when blockchain addresses are not even defined in the preceding contracts in such a scenario. The blockchain automatically resolves the dilemma by adding the parties' blockchain address to the subcontracts within the arbitration phase. This ensures that all parties involved in the dispute of each subcontract are easily made aware of the claims offered in the dispute. We understand the complexity of this procedure in our work and the computation cost involved in it. We thus provide a further means by which these contracts can be rearranged with the main contract to reduce the complexity involved. This will provide an effective way for contracts to be consolidated in situations where they become redundant in each dispute resolution.

A contract management procedure is designed to contain and manage the contract for each arbitration phase created. A sub-arbitration call procedure SAB_i is also created to provide means for sub-arbitration in a single pending arbitration. An evidence collection mechanism is designed to receive data from multiple parties. Connection is made with decentralized Apis, where analysis is made using decentralized oracles to ensure the correctness of data. From data received and evidence submitted, analysis is rendered on the blockchain. A decision procedure is created where rewards and penalties are created for arbitration by all parties [20, 21].

8.4 Sub-Contract Procedure Mechanisms

8.4.1 Contract-Contract Consolidation

All contracts under claim Cr_i to Cr_n are grouped as a set of contested contracts C_D to reduce redundancy among contracts collected by the blockchain to be disputed in each arbitration phase. A Contract Ranking Contract groups organizes these contracts in a hierarchical order. The contracts are ranked by the date they were created, their dependency, their relevance, and their reference. This ensures effective monitoring of all contracts set under the arbitration process. The arbitration phase $Abphase_i$ is given a transaction hash H and a blockchain address for all procedure calls SAB_n when it is formed on the blockchain.

For each arbitration process, all contract addresses are likewise stored on the blockchain [13]. The disagreement claims Cl_i are also documented, as well as the details of the parties involved. The transparency of the blockchain makes it easier to monitor transactions that take place in the arbitration phase [15] using the information recorded from the interactions among the contracts, dispute parties, and the claims. The rearrangement of the contracts to prepare for the arbitration phase makes it easier to work with each unique contract. It provides a smooth arbitration process to reduce the computation cost to a cost of O_n.

The contract list is saved as a DisputeContract Set DSC_i on the blockchain. This effective rearrangement of the contracts under dispute also provides a means to be able to contain new claims during a pending arbitration. For instance, in a dispute between an unforeseen contract C_i between two parties already involved in an arbitration phase $Abphase_i$. For each contract under dispute, a claim instance is established, and the information is added to the pending claim Arb_n.

8.4.2 Arbitration-Arbitration Consolidation

Our approach gives a better answer to this issue when two or more parties under a specific arbitration phase $Abphase_i$ are involved with the other entity in a bilateral dispute under a contract C discovered in both arbitration phases $Abphase_a$ and $Abphase_b$ operating concurrently. If a contract C_i is found in both existing arbitration phases, a party $Abphase_a$ might seek that the two arbitration phases be consolidated. This can only be done if all parties in $Abphase_a$ agree to transfer the contract to the other arbitration $Abphase_b$ to be challenged there, assuming no other parties in $Abphase_a$ are harmed. This is something they ask for on the blockchain. It is immediately launched on the blockchain if both parties agree and meet the pre-defined smart contract requirements.

Nevertheless, this solution may prove a challenge since, for each arbitration phase, there is a token of payment made before each arbitration phase is instantiated and a reward mechanism for each phase. Therefore, it will be uneconomical for party $Abphase_a$ to consolidate both arbitration phases together and lose an eminent reward intended to provide efficient management of contracts. We, therefore, provide a means for the reward and token to be kept under the new arbitration phase C consolidated for both processes. Once they are consolidated, a functional procedure in both arbitration and dispute resolution claims and clauses specified as functions in each smart contract are kept. Fees are therefore not required for the consolidation of arbitrations.

Furthermore, even if the contracts C_j are combined, the bilateral agreement between parties associated under a subcontract C_i is remains in effect. The prize for contract C_j in arbitration phase $Abphase_b$ is transferred to arbitration phase $Abphase_a$. This is combined with the dispute reward for the set of contracts that are being contested in the arbitration phase. Using game theory models, we can prove that there is fairness achieved from reward in a previous arbitration phase and in the new arbitration phase. However, we will leave this for future works.

8.4.3 Sub-Arbitration

In the blockchain networks, arbitration transactions are added as blocks of the network. Each transaction is hashed to the previous block transaction. This means that the first arbitration hash transaction is added as a subtree of the second arbitration hash creating interlinking set of nodes that is traceable on the blockchain. The capability for smart contracts to give off their rewards and payoff offers a very important advantage for the blockchain. Their interlinked capability with many other contracts also makes them relevant for managing contracts within an organization. The distribution of rewards can also be monitored all across multiple parties as rewards are passed from one party to the other.

For instance, since all parties will like to reduce their risk on claims, party employer E_i will shed responsibility on OutsourcedContractor OC_i who will also shed responsibility on *SubContractor SUBC_i*. Likewise, in rewarding parties, rewards can be received by the entity OC_i who has to make a compensation to party $SUBC_i$.

Effective distribution of rewards is observed, which ensures fairness for all parties involved. If the employer is awarded compensation, the outsourced contractor will like to pass on his liability toward the employer to the subcontractor $SUBC_i$ who ultimately caused the defects. However, instead of doing this in another arbitration phase, our scheme presents the best means by which the liability can be passed down to the subcontractor. All he has to do is invite the subcontractor to the arbitration process without instantiating another arbitration setup. This ensures a total minimization of arbitration creation, ensuring a reduction in computation and transaction cost. Once the subcontractor is invited, all procedure calls toward the subcontract dispute are likewise instantiated under the same arbitration phase $Abphase_i$.

8.4.4 Invitation

In the contract, a set of requirements Req is set on the blockchain. This provides a means by which a particular party can be invited to enter into an ongoing arbitration procedure $Arbp_{bi}$. The procedure call is made on the blockchain once a requirement procedure Req is set in each contract. This event call is used to invite a different entity i with a certain blockchain address to join an ongoing arbitration Ar_b. The blockchain validates the set of requirements Req to allow access to invitee I when a request is made to the party. He or she might next offer evidence for the procedure claim Cl for which he or she has been summoned. Thus, our proposed mechanism ensures an efficient means of managing arbitration and ensures a faster means of providing evidence reducing the time and cost of managing proceedings.

8.4.5 Dispute Negotiation

Dispute negotiation takes place among multiple parties in order to reach an agreement. A negotiation session $Negotiation_{ID}$ is initiated upon request by the parties on the blockchain. The phase for the negotiation $Negotiation_{phase}$ is set to 0. The claims Cl_i is made by disputing parties participating in the specific $Negotiation_{ID}$. All participating parties on the blockchain submit proposals on the blockchain over the contracts being disputed. If all parties are satisfied with the requests made, they make a decision. A time frame for the dispute scheduling period is set at which the decision is supposed to be made. Otherwise, the arbitration moves to another phase.

In ensuring that all parties effectively make decisions, our blockchain framework ensures that employees are systematically provided with the needed information on matters that concern them through the transparency of the blockchain. If all the documents and all evidence have not been provided, each party can request a reschedule. Time t is extended based on specifications made in the smart contract. However, we provide an allowance of two reschedules in a smart contract. Also, in order to prevent a denial-of-service attack, fees are collected from the party requesting a rescheduling. Once all the information is processed and analyzed on the blockchain. We call this machine-based arbitration.

8.4.6 Blockchain-Based Arbitrators

Blockchain-based arbitration processors serve as the arbitrators for each dispute. The blockchain-based arbitration processors are made up of a group of decentralized arbitration nodes implementing analytic algorithms. They are randomly chosen to execute the analytical algorithm without the help of external parties. Due to our blockchain-based analytic systems' independent nature, delays are avoided with multiple arbitrations since decisions are made quickly, and the blockchain provides a means for decisions to be made quickly. None of the parties is responsible for selecting a blockchain-based arbitrator, ensuring the independence of the system.

Hence, all parties have a large degree of fairness [24]. The blockchain-based arbitration processor analyzes the contracts given. The smart contract is binary and is less ambiguous. They are, therefore, more certain. Transactions are therefore cheaper and code enforceability easier. However, to ensure effective interpretation of contracts designed, the smart contracts are combined with plaintext natural language to provide a straightforward interpretation for all parties involved.

8.4.7 Arbitration Appeal

In complex cases where a claim is decided and a reward is paid to Party A, a term is set during which Party B might submit a new dispute claim C_d. During this time, the smart contract holds the claim, and if $t(x) > t$ (the time limit for filing an appeal I has passed), the *ArbitrationContract* transfers the claim reward to party A; otherwise, the reward is held until the second arbitration decision is made before the reward from arbitration Ar_b is distributed.

8.5 Internal Contracting Considerations

8.5.1 Security

Managing multi-party arbitration using the blockchain helps prevent contracts from being tampered with in-dispute resolutions. Multi-party arbitration in each organization helps maintain the smart contract's sustainability and promotes efficiency in the dispute resolution process. The blockchain facilitates ease in contract calls. The connection between multiple contracts is easy to trace from various contract addresses and access-control functions, and security is ensured between these contracts being used in multiple organizations.

8.5.2 General Contract Design

Each contract is written and encoded on the blockchain as a smart contract. The contracts designed are deployed and allocated blockchain addresses as *Add*. The smart contracts provide a set of programming logic from which the drafted set of contracts is converted into enforceable executable logic. This encapsulates the set of agreements between the various parties. Although automated, with a set of logic drafted without any intermediary to ensure its enforcement, our smart contract developed in this work satisfies both common and contract law conditions because it provides a means of promise and guarantees the enforcement of the rules.

Once the contract is deployed, the particular contract C_i usable for negotiations must be approved by all network parties. This approval action is based on all parties' signatures recorded and stored on the blockchain. For each smart contract deployed on the blockchain, a requirement is set up on the blockchain, which specifies each contract's rules between the employer and the employee. The job descriptions, offerings, and payments are specified in the contract. Blockchain addresses are assigned to each contract and every interlinked contract. Rewards and penalties are specified in the contract upon breach of the contract created.

8.5.3 Contract Termination

If the contract deployment circumstances must be changed, the contract between them must be canceled. To terminate the contract in use, all parties on the network pay a charge. The contract can be dissolved once all parties have paid the amount. However, the contract must be dissolved with the approval of all participants via a multi-signature mechanism. The contract status changes to terminated after the approval is finalized. However, the contract still remains recorded and stored on the blockchain due to the blockchain's immutability. Since the contract also has access to the multiple parties' blockchain addresses, the contract still remains valid even after approval.

However, since a contract is an agreement between participants, there has to be a new means to replace it. Once a contract is terminated, it is still linked to other contracts under it. This makes it a highly complex process to terminate contracts.

In our scheme, when a contract is terminated, we send out a notification to all blockchain network members, as well as the contracts affected by the termination and their addresses, to ensure that these contracts are updated and taken care of. In its stead, a new contract C_2 has been deployed. A new contract is deployed that includes a reference to the prior contract C_1 as well as references to other contracts that are reliant on it. Thus, in the event of situational changes, we provide a framework by which the contract set can be terminated by the approval of all parties involved and not by the individual's approval. This also makes making a change to a contract already set more difficult.

In the contracting design, we offer a framework that must be looked at in the construction of a blockchain-based contract for each organization in ensuring dispute resolution on the blockchain.

- On the blockchain, a dispute D_i must be reported.
- Each dispute D_i must include at least two participants, $M > 1$, where M denotes to D_i members.
- Contracts C_{main} that are under dispute D_i must be assigned.
- Subcontracts must be set for C for D_i as C_{sub}.
- Arbitration procedures for all C_{main} and C_{sub} are called on the blockchain.
- Unique IDs for contracts associated with each arbitration Ar_b process must be supplied.
- There must be a dispute D_i claim Cl under the main arbitration phase $Abphase_{main}$ for a sub-arbitration Sub_{arb} to be created.
- Sub-arbitration process created Sub_{arb} must be unique with a unique blockchain hash.
- All parties who have added their blockchain addresses *add* must participate in the main arbitration phase Ar_b.
- For every sub-arbitration process produced during the main arbitration phase, sub-arbitration procedures Sub_{arb} are a must-have reference to main arbitration procedures Ar_b.
- Each invitee I to each arbitration phase Ar_b must meet the requirement Req.
- The previous arbitration stage's decisions Dec and award R_{ewd} must be referenced to a newly established sub-arbitration Sub_{arb} with references to the parties' addresses *add*.

8.6 Contract Analysis

8.6.1 Post-Facto

This is used to describe events and settings that change after contracts are deployed. Such situations need to be accounted for. Interpretation for all new situations must thus be able to be consistent with smart contract deployed and seen there same by all agreeing members. We, therefore, account for ex-post negotiation on the blockchain after the contract has been drafted between all parties involved. In this case, all parties agree to enter a post-facto state. Each contract put on the blockchain includes the option for a post-facto. One participant makes a request to all other parties in order to start a past-facto request. The prior contract is terminated and a new one is executed if all parties agree. The essential changes that have occurred are accounted for in this contract, which were not accounted for in the prior contract.

All parties consent to their secret keys, and a new contract is created and deployed on the blockchain network by all parties. If the post-facto is denied, a new arbitration process is started to address the situation. In contract drafting, the plaintext provides a proper interpretation and ensures a practical judgment for the next contract C_2 deployed over C_2.

8.6.2 Interaction between Contracting Parties

We will look at a contract negotiation between an employee Emp_a and an employer Emp_b who are working on a project in an organization that calls for action a, among other things $A = \left[a^1, a^2\right]$. Each party i has an initial power of control p, a utility level B_i, and receives a private benefit B_i from the chosen action a through the agreement B_i. Each party has the utility $u_i = b_i + m_i$ to obtain the benefits under this contract, where m signifies the net power of i to do so. The following assumptions essentially capture the potential conflict of interest between the two parties:

1. *Assumption 1*: In a, b_1, and b_2 are constantly increasing and decreasing. Assuming that party i's chosen action is a, we can see that b_1 and b_2 are both concave in a, with the benefits of these actions increasing and decreasing as players make decisions. Each person strives to maximize his or her own benefit from the deal. A Pareto-efficient action will occur at some time during the game, resulting in a Pareto-optimal profit for all participants. This is referred to as $a^* \equiv \arg\max\left(b_1(a) + b_2(a)\right)$. A degree of productivity P is gained by allowing the a^* to be played on the blockchain.

2. *Assumption 2*: If $A2 : b_1(a^*) + b_2(a^*) > B1 + B2$ then Assumption A1 suggests that $b_1^1 > b_1^2$ and $b_2^2 > b_2^2$. We examine instances in which the ex-post cannot be verified by the blockchain. We suppose that a specific period t will be required to attain the Pareto-optimal action a^*, which will offer the required efficiency, because not all of the actions made by employees in the smart contract are efficient. However, there is a high chance P that each party can exploit. Even if token transfers are incentivized to determine how modifications are made, there is still a high likelihood of exploitation. The economic climate, frequency of interactions, reputation and credibility, and other unforeseen facts that can create a change in the already established

contracts between the two parties will all influence the extent to which parties can reach a more efficient agreement. When the optimal action a^* is found, the advantages to each of the participants will outweigh the benefits of a set of acts performed without the blockchain. The blockchain does not make contracting between parties not only beneficial, still, it also makes it easier for future actions not to be altered and provides an interface for contracts between all parties to collaborate toward enhancing the security of the established contract in a peer-to-peer manner. The benefits are still assured, as expressed in Assumption 1.

8.6.3 Allocating Decision Rights between Contracting Parties

The topic of assigning decision rights to parties across a blockchain network will be discussed. However, in our instance, we're looking at the collaboration between two parties [1, 2]. The decision rights are safely managed through a mechanism on the blockchain that neither party does not influence. Thus, a fair game is created in the organization for each party to engage in negotiation.

At the ex-post negotiation commencement, an initial proposal is sent across the network to all parties involved in the negotiation. This is the first transfer before additional $a_i, \ldots, a_n$ negotiations. The equitable distribution of negotiations will be enabled by allocating future decision rights based on recorded activities based on the transparency of the blockchain. The right of each party to make their own decision has no bearing on their bargaining. It just establishes the method by which the game is played [1, 2] by the party doing action a_i and making decisions.

The most important aspect is to watch how the party with the decision acts from the beginning $(a, t = t_0)$ to the end $(a, t = t_n)$. The generalized Nash-bargaining solution with bargaining powers α_1 and $\alpha_2 (\alpha_1 \geq 0, \alpha_1 + \alpha_2 = 1)$ determines the final conclusion of the negotiation, providing that the set of transfers allowed in each proposal is limited to a limit T. The set of decisions that will be made in the end is determined by

$$= \max_{a \in A, t} \left[b_1(a) - t - \left(b_1(a \wedge) - t \wedge \right) \right]_1^\alpha$$
$$\left[b_2(a \wedge) + t - \left(b_2(a \wedge) + t \wedge \right) \right]_2^\alpha$$

A limit will ensure a specified point at which a decision can be taken, as well as the outcome of the blockchain network. As a result, with the help of the blockchain, the ideal negotiation strategy that benefits all stakeholders inside the corporation can be selected. Each new contract C_1 negotiated must have the parties' signatures, control rights, the initial transfer, and an ex-post mechanism to be used in the event of future events. In the event of an unforeseen circumstance, the eventual acts become easier to negotiate.

8.6.4 No Leverage Constraints between Both Party A and Party B

When both parties A and B have no leverage limits, such as in a payment discussion between an employer E_1 and an employee Emp_1 within an organization O, the negotiating power is enormous. With this circumstance, we are still aiming for ex-post efficiency in the smart contract we've created. This also implies that, regardless of who owns the smart contract's original distribution of decision powers, an efficient action $a*$ must be taken. However, whoever has decision-making authority can use this to seek several renegotiations. If the proposed original contract is signed (δ, t_0) in the first step, the renegotiation conclusion will be the same and is thus defined by $a = a*$ and a net transfer, as follows:

$$\left[b_1^* - t - \left(b_1^\delta - t_0 \right) \right] = \alpha \left[b_2^* + t - \left(b^\delta - t_0 \right) \right]$$

The final levels of utility between Party A and Party B are equal to $u_1(\delta, t_0)$ and $u_2(\delta, t_0)$, which results to

$$u_1(\delta, t_0) = b_1^\delta + \alpha_1 (b_* - b_\delta) - t_0 \tag{8.4}$$

$$u_2(\delta, t_0) = b_2^\delta + \alpha_2 (b_* - b_\delta) - t_0 \tag{8.5}$$

Because the consequent advantage makes the contract agreement viable according to Assumption 2, each δ in 1, 2 results in a transfer $t_0(\delta)$ in the contract negotiations.

8.6.5 Leverage Constraints

There may be a difference in leverage between the parties to a contract that can be renegotiated. Therefore, we make it impossible for parties to renegotiate established contracts using their leverage (w). So we stop parties from misusing the contract. Because all parties to the contract stand to gain financially and personally, giving both parties decision control without the correct condition is not sustainable. We also imposed a transfer cap. However, limiting transfers between both parties is problematic because renegotiating is intended to achieve ex-post efficiency. If a limit T is required, it is stated as

$$w_1 + w_2 < W \equiv \alpha_1 (b_2 * 2 - b_2 * 1) - \alpha (b_1^* - b_1^1)$$

Proof

If the theorem is correct, ex-post efficiency cannot be achieved by allocating the employee's decision correctly and then setting the initial transfer to the employer. Similarly, delegating control to the employer will not allow us to achieve ex-post efficiency if we have $u_1^1 - u_2^2 = u_1^2 - u_1^2 = w_2 - w_1$

As a result, the most efficient and acceptable contract is one in which $(\delta = 2, \rho = -w^2)$. That is, a smaller initial transfer (in absolute value) $t_0 > -w^2$ initially boosts the employer's utility but decreases both the employee's utility and efficiency over time t_n.

8.7 Conclusion

This chapter looked at the interaction between contracting parties over a blockchain network. The focus is on multi-organizational contracting among these organizations and sub-units located across multiple organizations; in the work, parties transmitted transactions across various networks, visible on the blockchain utilizing contracting frameworks defined. We offered novel mechanisms for approaching contract definition and multi-contracting in recruitment, outsourcing, sub-outsourcing, contract arrangement, and arbitration. We also developed a more adaptive framework utilizing blockchain smart contracts. Our results proved the efficiency of our proposed solution.

References

[1] Belanche, D., Casaló, L. V., & Orús, C. (2016). City attachment and use of urban services: benefits for smart cities. Cities.

[2] Kim, T. H., Kumar, G., Saha, R., Rai, M. K., Buchanan, W. J., Thomas, R., & Alazab, M. (2020). A privacy-preserving distributed ledger framework for global human resource record management: the blockchain aspect. IEEE Access, 8, 96455–96467.

[3] Onik, M. M. H., Miraz, M. H., & Kim, C. S. (2018, April). A recruitment and human resource management technique using blockchain technology for industry 4.0. In Smart Cities Symposium 2018 (pp. 1–6). IET.

[4] Sifah, E. B., Xia, H., Cobblah, C. N. A., Xia, Q., Gao, J., & Du, X. (2020). BEMPAS: a decentralized employee performance assessment system based on blockchain for smart city governance. IEEE Access, 8, 99528–99539.

[5] Pokrovskaia, N. N., Spivak, V. A., & Snisarenko, S. O. (2018, November). Developing global qualification-competencies ledger on blockchain platform. In 2018 XVII Russian Scientific and Practical Conference on Planning and Teaching Engineering Staff for the Industrial and Economic Complex of the Region (PTES) (pp. 209–212). IEEE.

[6] Webb, J. N. (2007). Game theory: decisions, interaction and evolution. Springer Science & Business Media.

[7] Neiheiser, R., Inácio, G., Rech, L., & Fraga, J. (2020). HRM smart contracts on the blockchain: emulated vs native. Cluster Computing, 1–18.

[8] Carlsten, M., Kalodner, H., Weinberg, S. M., & Narayanan, A. (2016, October). On the instability of Bitcoin without the block reward. In Proceedings of the 2016 ACM SIGSAC Conference on Computer and Communications Security (pp. 154–167).

[9] Brousseau, E., & Glachant, J. M. (Eds.). (2002). The economics of contracts: theories and applications. Cambridge University Press.

[10] Cvitanic, J., & Zhang, J. (2012). Contract theory in continuous-time models. Springer Science & Business Media.

[11] Prem, C. (2020). The theory of credit contracts. Springer.

[12] Zheng, Z., Xie, S., Dai, H. N., Chen, W., Chen, X., Weng, J., & Imran, M. (2020). An overview on smart contracts: Challenges, advances and platforms. Future Generation Computer Systems, 105, 475–491.

[13] Dannen, C. (2017). Introducing Ethereum and solidity (Vol. 1, pp. 159–160). Apress.

[14] Benson, E. (1988). The law of industrial conflict. Springer.

[15] Schmitz, A., & Rule, C. (2019). Online dispute resolution for smart contracts. Journal of Dispute Resolution, 103.

[16] Gudkov, A. (2020). Crowd arbitration: Blockchain dispute resolution. Legal Issues in the Digital Age. https://digitalawjournal.hse.ru/article/view/11780.

[17] Wing, L., Martinez, J., Katsh, E., & Rule, C. (2021). Designing ethical online dispute resolution systems: the rise of the fourth party. Negotiation Journal, 37(1), 49–64.

[18] Goldenfein, J., & Leiter, A. (2018). Legal engineering on the blockchain: 'Smart contracts' as legal conduct. Law and Critique, 29(2), 141–149.

[19] Ortolani, P. (2016). Self-enforcing online dispute resolution: lessons from Bitcoin. Oxford Journal of Legal Studies, 36(3), 595–629.

[20] Kondev, D. (2017). Multi-party and multi-contract arbitration in the construction industry. John Wiley & Sons.

[21] Schmitz, A. J. (2020). Making smart contracts "smarter" with arbitration. American Arbitration Association website.

[22] Frith, J. (2017). Big data, technical communication, and the smart city. Journal of Business and Technical Communication, 31(2), 168–187.

[23] Castro, M., Jara, A. J., & Skarmeta, A. F. G. (2013). Smart lighting solutions for smart cities. In Proceedings – 27th International Conference on Advanced Information Networking and Applications Workshops, WAINA, 2013.

[24] Kaal, W. A., & Calcaterra, C. (2017). Crypto transaction dispute resolution. The Business Lawyer, 73(1), 109–152.

[25] Hasen, R. L. (2014). Examples & explanations for remedies. Wolters Kluwer Law & Business.

[26] Farnsworth, E. A. (1970). Legal remedies for breach of contract. Columbia Law Review, 70(7), 1145–1216.

[27] Benligiray, B., Connor, D., Tate, A. and Vänttinen, H. (2019). Honeycomb: An Ecosystem Hub for Decentralized Oracle Networks. [online] Available at: https://www.clcg.io/whitepaper/

[28] BlockChainHub. (2019). Types of blockchains & DLTs (Distributed Ledger Technologies). https://blockchainhub.net/blockchains-and-distributed-ledger-technologies-in-general/.

[29] Chourabi, H., Nam, T., Walker, S., Gil-Garcia, J. R., Mellouli, S., Nahon, K., … & Scholl, H. J. (2012, January). Understanding smart cities: an integrative framework. In 2012 45th Hawaii International Conference on System Sciences (pp. 2289–2297). IEEE.

[30] Gascó-Hernandez, M. (2018). Building a smart city: lessons from Barcelona. Communications of the ACM, 61(4), 50–57.

[31] Li, S. (2018, August). Application of blockchain technology in smart city infrastructure. In 2018 IEEE International Conference on Smart Internet of Things (SmartIoT) (pp. 276–2766). IEEE.

9

Smart Energy

9.1 Introduction

Currently, the utilization of power is seen as the basic supporting component that empowers the revelation or advancement of innovations [1]. Power as a product has become the most important driver of innovation, and most ideas are restricted or worthless without it [2, 3]. Its application has spawned and fostered advancements in a variety of human endeavors. Transportation, communication, computers, and other fields have all benefited from increased electricity. Its portability and widespread use have piqued the curiosity of various research institutes, prompting them to investigate new methods for ensuring its efficient delivery and testing. One result of such interests is the smart grid [4].

Smart grid technology allows utility companies and their customers to communicate in two directions. It supports the development of plug-in charging for electric vehicles by allowing for greater integration of more current energy generation technology such as wind and solar-powered electricity [5]. By logging and sending data to the smart grid via a smart meter, the smart home interacts with the intelligent network [6, 7]. Customers can better manage their energy consumption this way. Utility providers [8] can supply their customers with greater data to monitor and educate how much they pay for the service [9] by more accurately assessing a home's power usage levels with a smart meter.

Power usage data from smart homes is transmitted to the smart grid network in real time over the Internet, where it is analyzed and stored in databases for charging consumers and research purposes by various research organizations [10, 11]. When hostile actors gain access to the information that is recorded and transferred with the smart grid network via Internet, it can be hacked. When such data is compromised, the customer is frequently forced or compelled to pay quantities that are insufficient to compensate for the benefit received at the conclusion of the charge term, which is usually monthly [12]. This could result in the client overpaying or the power provider losing money. In addition to the aforementioned issue, consumers are not informed about charging points, so they are unaware of which

DOI: 10.1201/9781003289418-9

household devices consume the most electricity [13], data that might be used to educate purchasers and optimize the use of such products to save expenses [13, 14].

Several solutions have been presented to overcome the challenges that have arisen as a result of the use of smart grids in recent years. Fan et al. [15] highlighted the difficulties with today's electricity distribution frameworks. They discovered a number of flaws in current power distribution frameworks, including a lack of computerized inspection, inadequate visibility, mechanical switches affecting average reaction times, and a lack of situational mindfulness, to name a few. They went on to suggest the smart grid as a solution to these problems. The various communication advancements associated with the smart grid were also discussed. ZigBee, wireless mesh, cellular network transmission, power-line communication, and enhanced supporter lines are examples of these.

Güngör et al. [16] presented a Cloud computing paradigm for monitoring real-time smart grid data streams. Their concept serves as a platform for collaboration and data exchange among clients, retailers, virtual control plant managers, and network administrators. Their focus was on the Cloud computing characteristics that led to an Internet-scale infrastructure that can support the smart grid's data-intensive requirements. Rusitschka et al. [17] suggested a smart grid security innovation based on public key infrastructure (PKI) technologies, which uses advanced certificates to link open keys to client identities (IDs). They advocated that PKI guidelines be improved for use by the fundamental foundation industry. These standards will be used to establish requirements for energy service companies' PKI operations.

For smart grid adaptability, Metke et al. [18] proposed a blockchain. This concept employs blockchain and smart contracts as mediators between power customers and producers in order to reduce costs, speed up exchanges, and strengthen the security of the data created. Whenever a communication happens, a blockchain-based meter updates the blockchain by creating a unique timestamp piece for confirmation in a distributed ledger, according to their methodology. System administrators charge customers based on the information recorded on the blockchain at the conveyance level.

Mylrea et al [19] suggested an ID-based security solution to allow utility firms to have the most security control over the data related while also allowing the Information and Communication Technology (ICT) system to manage more customers. An identity-based signcryption (IBSC) scheme lies at the heart of this scheme. At the same time, this plot serves the functions of digital signature and encryption. Their method employs public key cryptography, which is computed based on each client's ID and a set time limit, beyond which the framework is unable to compute the public key.

Nonetheless, the blockchain might be viewed as a technological advancement that can provide a reasonable solution to the challenges raised by its attractive properties of immutability, non-repudiation, and decentralization. This study presents a blockchain-based approach that is combined with smart contracts to create a tamper-proof framework for guaranteeing buyer information is recorded and shared on the smart grid framework. The blockchain will be used to establish an immutable data structure in which data stored and exchanged on the platform cannot be changed. The smart contract will be used to establish rules between customers and utility businesses, as well as to show algorithmic contract terms in the event that any substance is deemed to be illegal.

9.2 Preliminaries

This section discusses the technologies and ideas that are needed to establish concrete understanding and designing of the secured sovereign blockchain-based monitoring on the smart grid.

9.2.1 Blockchain

The blockchain is a distributed ledger that contains an orderly arrangement of entries connected by blocks and chains. Individual elements of the blockchain, known as blocks, contain data on a particular transaction between entities. Logs of a single occurrence are a good example of data in a block. A blockchain network keeps track of an ever-growing list of irrefutable records. This capability enables systems based on sovereign blockchain technology to distribute assets securely among mutually untrusted entities.

The processing and consensus nodes are solely responsible for converting transactions into blocks and broadcasting blocks onto the blockchain network in the smart grid's blockchain-based monitoring. Forms are generated by the processing and consensus nodes. Any event that is transferred into the blockchain network, formed into blocks, and then broadcast on the blockchain network is covered by these forms. The activity outlined earlier completes a partnership and permits a block to be broadcast into the blockchain network. In the network, there are several blockchain threads, each of which is identifiable by a consumer's ID. Threading side blocks to their parent blocks creates a continuous log of well-ordered logs based on requests from various consumers.

Structuring the network this way reveals that each block in a particular string represents different instances of events. The smart contracts are entered and modified in a side block that is attached to the parent block. The relevance of integrating side blocks is to maintain an effective log and efficient block fetching, with a focus on querying and investigation in the event of a breach of agreements by consumers and utility providers.

The content of the side blocks appended to their parent blocks are smart contract reports that are keyed with users' IDs, ensuring accuracy on identical pieces with violations stored in the smart contract database at the same time. The structure of generating many threads for the sovereign blockchain network, as well as generating side blocks on parent blocks, results in a large number of reports.

A block is created from a form established by a transaction between entities, as indicated before in this section. A single transaction makes up a whole block, from the time it is formed to the time it is made available on the smart grid system. Only the processing and consensus nodes have full access to the sovereign blockchain network. Nodes monitor the blocks, including parent and side blocks, and inform the system when the agreed-upon use of data is violated.

9.2.2 Smart Meter

A smart meter is a new form of electric and gas meter that employs wireless technology to digitally transfer meter readings from consumers' houses to energy suppliers in real time, resulting in more correct energy bills. In other terms, a smart meter is an electronic gadget that reads and records electric energy use and communicates that data to the utility provider for tracking and invoicing. The smart meter includes an in-home display screen that displays how much energy a user uses in real time.

Between the meter and the energy supplier, there are two-way communication connections. Wireless contact between the meter and the energy provider is possible. ZigBee (low-power, low-data-rate wireless) and Wi-SUN are two examples (smart utility networks). Fixed wired connections can also be used. The power-line carrier is an example of a fixed wiring link.

A smart meter is used to digitally transfer meter readings from a consumer's residence to the sovereign blockchain network in this project. The smart meter readings are converted into blocks and then added to the sovereign blockchain after being confirmed and accepted by the majority of processing and consensus nodes. Between the smart meter and the sovereign blockchain network, smart contracts protocol is also built. The smart contracts are triggered and executed based on the observed activity, if illegal behavior is present on the meter, and the required action is done.

9.2.3 Internet of Things

The Internet of things (IoT) has ended up the subject of the day as far as innovation, industry, policy, and engineering circles are concerned. IoT is epitomized in a wide range of organized items, frameworks, and sensors, which take advantage of progressions in computing control, hardware miniaturization and interconnections to offer modern capabilities not already conceivable. Right now, the innovation is being executed on an expansive scale, which has contributed to changing the way we live. Consumers are presently displayed with new IoT gadgets like Internet-enabled machines, domestic mechanization components and vitality administration gadgets. These gadgets are moving us toward a smart domestic vision, advertising more security and vitality proficiency.

IoT can be characterized as a network of physical gadgets, vehicles, domestic machines and other things embedded with electronics, software, sensors, actuators, and connectivity. These objects are given unique identifiers and the capacity to exchange information over a network without requiring human-to-human or human-to-computer interference. The thing within the IoT can be named an individual with a heart monitor implant, a farm animal with a biochip transponder, a built-in vehicle sensor to caution the driver when tire weight is low. The basic highlights of the IoT are artificial intelligence, connectivity, sensors, dynamic engagement, and little gadget utilize.

The IoT has spawned personal devices such as wearable fitness, health monitoring devices, and network-enabled medical equipment, all of which are revolutionizing the way healthcare is given to people. The good news is that IoT benefits those with disabilities and the elderly since innovative technologies have enhanced their degree of independence and quality of life at a fair cost. IoT has enabled systems such as network cars, sensors, and intelligent traffic systems to be installed in our roads and bridges in recent years. These solutions are bringing us closer to smart cities, which assist reduce traffic and energy consumption. We now have the opportunity to alter our agricultural, industrial, and energy production by expanding the availability of information provided along the value chain of production via networked sensors, thanks to the arrival of IoT.

When used in governance and safety, the IoT improves law enforcement, defense, municipal planning, and economic management. We have been able to fill in the gaps, fix several problems, and broaden the scope of our activities thanks to the IoT. For instance, with IoT, city planners [20] can see the influence of their designs more clearly, and governments can better comprehend the state of the economy. IoT functions in marketing and content distribution in a way that is analogous to current technology, analytics, and Big Data. Current technology allows for the collection of specific data in order to generate associated metrics and patterns across time. The information gathered is frequently shallow and inaccurate. The IoT, on the other hand, improves data collecting by seeing and analyzing more behaviors.

The IoT uses a variety of protocols and networking technologies, although RFID, NFC, low-energy Bluetooth, low-energy protocols, LTE-A, and WIFI-Direct are the most important enabling technologies and protocols. In contrast to a conventional uniform network of common systems, these technologies aid network functionality, which is required in IoT systems.

9.2.4 Smart Contracts

Smart contracts are pre-programmed operations that are engaged and executed when an action is received. These agreements are made between the parties engaged in a certain transaction. The agreements found there are spread throughout a decentralized blockchain network. Smart contracts, in general, allow multiple, anonymous parties to conduct trustworthy transactions without the need for a central authority, legal system, or external enforcement mechanism. Smart contracts are traceable, irrevocable, and transparent in systems that use them.

The smart contracts on this system are set up to automatically send reports generated by action activation. The major purpose of smart contracts is to detect malicious use of electrical power and data and to report such behavior to a database. When a consumer uses electrical power maliciously, the smart contracts terminate that consumer's access to electricity. When malicious manipulation of electrical data about a user is detected, the smart contracts notify the user.

9.2.5 Smart Grid vs. Traditional Grids

Traditional power grids had nearly no storage capabilities; they are demand-driven and have a progressive structure. In a power network, voltage is continuously brought down so these diverse buyers can utilize the power: from transmission voltage levels to conveyance voltage levels to service voltage levels (in reality, it's equipping up and down and, in this way, a bit more complex). Regularly, a distinction is made between transmission and dissemination, where distinctive wiring and cabling frameworks come within the picture. An electrical grid is to make beyond any doubt that power is continuously given when and where required, without intrusion – and in this lie numerous challenges where a smart grid can already offer solutions. Given the complexity, the numerous challenges that can emerge, such as the results of serious climate conditions, harm by natural life, human disrupt and other outside components and inside factors overseeing a network is exceptionally complex and a committed field for specialists who moreover need to consider the choices concerning energy controls and activities by governments.

In smart grids, self-healing capabilities enable identifying and reacting to network issues and guaranteeing fast recovery after disturbances. The two-way stream of power and information that's the fundamental characteristic of a smart grid enables to feed data and information to the various partners within the power market that can be analyzed to optimize the lattice, anticipate potential issues, respond speedier when challenges emerge and construct modern capacities and administrations as the power scene is changing. The power market, the utilization of power, regulations, requests of different partners, and the exceptional generation of power are all changing. So, intelligent grid activities exist over the globe, though in some cases with diverse approaches and objectives. Whereas smart grid still alludes to the bi-directional transmission of information and power (with prosumers and organizations creating power as well), the meaning and reach of the term has broadened given the numerous conceivable outcomes empowered by this imperative alter and ever more advances utilized in a setting of smart grid deployments.

This incorporates, as already said, IoT technologies, Big Data and progressed analytics with artificial intelligence and machine learning on top, sufficient communication benchmarks utilized to send information from one point to another and more innovations that we see popping up within the advanced change of utilities and in Industry.

9.2.6 Cryptographic Keys

Our solution uses cryptographic keys to perform system and data security tasks. We present the keys required to ensure blockchain-based smart grid monitoring between buyers and corresponding utility firms. These keys allow us to secure our system. Our system requires cryptographic primitives for the secure flow of data from the smart home to the smart grid to prevent hostile actors from interfering with or altering data in transit. We depict the primitives and keys that best suit our system, which include the following:

- *Consumer private key*: This is usually created by the consumer, who must carefully sign requests for data access.
- *Consumer public key*: This is a key generated by the consumer and delivered to the smart grid network's authenticator. It is used to verify the consumer's identification in order to gain access to information. The public key is also used to encrypt data that the authenticator sends to the consumer.
- *Authenticator contract key*: This is often a key pair generated by the authenticator and linked to a smart contract in a package that is used to encrypt reports sent from the consumer's system to the smart grid network and vice versa.

The consumer creates a key match (customer private key and public key), saves the private key, and publishes the public key with the smart grid. After that, the client creates requests, signs them with his private key, and transmits them to the smart grid network. The authenticator confirms the request by checking the signature with the open customer key after obtaining the information. The results of computations on the data sought are placed in labels included in the information, and the eligible businesses then prepare the data on the smart grid network.

The authenticator contract key is used to encrypt the processed data before it is transmitted to the consumer. The consumer decrypts the encrypted bundle and consumes the data as it arrives. The authenticator contract key tagged to the contracts, which the data originator has already generated, is used to adequately secure the exchange of data from the consumer to the smart grid network, the detailing of activities and events created from the use of cryptographic capacities, and the flags raised on the data. As a result, this information is forwarded to a secure database.

9.3 System Design

This section provides an overview of our system's whole system design. We provide a high-level overview of the many layers that we used in our system. This section also explains the relationship between the various layers of our method. We also go through the data transfer mechanism we used for data security and transparency in our secured blockchain-based smart grid monitoring between consumers and utility corporations. Figure 9.1 depicts a high-level overview of our secure blockchain-based smart grid monitoring.

9.3.1 Initials

This part gives the meanings of the various initials used in this chapter.

- *TTPP*: Time to Purchase Power
- *TOC*: Time of connection
- *TTPT*: Time To process transaction
- *TOR*: Time of request
- *TTPR*: Time to process request
- *TPRT*: Time power reaches threshold value
- *TPF*: Time power finishes
- *MID*: Meter identification
- *HN*: House number
- *APP*: Amount of power purchased
- *NID*: Node identification
- *Nsig*: Node signature
- *TVLN*: Type of violation
- *TSM*: Timestamp of state of meter
- *TSV*: Timestamp of violation

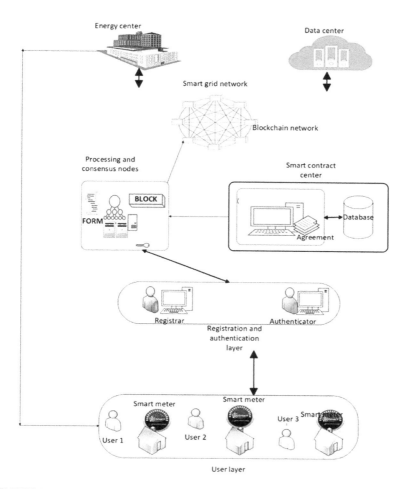

FIGURE 9.1
A general overview of our secured sovereign blockchain based monitoring on the smart grid.

The three primary layers of the secure blockchain-based smart grid monitoring system have been identified. The user layer, data processing and monitoring layer, and energy production and data storage layer are the three layers in question. The data processing and monitoring layer directly interface with both the user and energy production and data storage layers. The details of the three layers stated previously are explained in the following.

9.3.2 User Layer

The user layer is made up of all entities that use power from a certain utility company to start and run basic processes for their own purposes. This layer communicates directly with the data processing and monitoring layer's registration and authentication center. The user layer allows users to sign up for the system and become a member of it. The data processing and monitoring layer receives information about the system's members' registration. The registration and authentication center receives and processes it first. Individual users in homes or offices, schools, healthcare facilities, organizations, and other institutions are examples of system users.

9.3.3 Data Processing and Monitoring Layer

Individual components make up this layer, which aid in the processing of all data transmitted to the smart grid network. This layer also performs computations on the data and tags it with functions that aid in the monitoring of all actions carried out on the system. Algorithms are established on this layer to automatically report any illegal behaviors that occur in the system and to initiate an endeavor to prohibit access to electricity usage. The reported illegal behaviors are marked with the accompanying user's unique ID and saved safely in a database. Furthermore, every attempt sent to the system is broadcast over the network, ensuring that auditing is both trustworthy and fair. Finally, this layer is in charge of authenticating all system actions and requesting data access. The individual entities that live on this tier are discussed in further depth in the below sections.

9.3.4 Registration and Authentication Center

This layer includes the registrar and authenticator. When users register on the user layer, the registrar at the registration and authentication center obtains their credentials first. This information is processed by the registrar, and a unique ID for the matching user is generated. The identifier is the unique ID, which is used to identify each user on the system. Every user's unique ID is supplied to the system's authenticator. When a user connects in to the system, the authenticator verifies his ID using his unique ID.

9.3.5 Processing and Consensus Nodes

Consumer power usage data, including requests and access granted, is handled by the processing and consensus nodes. A processing and consensus node takes a consumer's energy request details, then adds the consumer's matching meter ID (MID), house number, and area code and processes these into a temporary form. After then, the form is turned into a block, which is sent to other nodes for verification and acceptance. The block is uploaded to the main sovereign blockchain after it has been confirmed and accepted by the majority of processing and consensus nodes. If it is not accepted, however, it is returned to the originator for reconsideration. The form is copied and transmitted to the data center for storage.

For forecasting and research, the utility company uses data from the data center and research agencies.

9.3.6 Smart Contract Center

The smart contracts used in this study function as a finite state machine, executing the rules agreed upon by utility companies and consumers. When the smart contracts receive a trigger, the rules are immediately enforced, and the offender is sanctioned appropriately. When a trigger is received, the smart contracts also send reports from the smart homes to the smart grid network. Furthermore, malicious acts on data received onto the smart grid network via the smart meter are automatically reported by the smart contract. Consumers will be able to acquire assurance and control over data provenance as the full lifecycle of data supplied to the smart grid network will be monitored, regulated, and stored in a secure environment. Smart contracts send reports to the smart grid network, which are first stored in a permissioned database, then retrieved by processing and consensus nodes and processed into blocks, which are then broadcast to the blockchain network.

9.3.7 Energy Production and Data Storage Center

This layer connects to the data processing and monitoring layer directly. The layer in question is divided into two portions. These are the data storage center and the energy production center. The energy center is in charge of generating electrical power and transferring it to various smart homes in response to requests from the data processing and monitoring layer's processing and consensus nodes. The generated energy is distributed to registered clients on the smart grid network according to monthly pricing. Records are taken and afterward added to the client's information before being turned into blocks. After the majority of processing and consensus nodes have verified these blocks, they are added to the main sovereign blockchain. The data center is a storage facility with connections to the data processing and monitoring layers. In other terms, it is a repository for forecasting and research that stores copies of consumer data that has been processed into blocks. Furthermore, because the data supplied to the data center is linked to the source of the data, it is monitored.

9.4 Data Flow on the Entire System Description

Users registered on the system have unique IDs tagged with a smart meter and installed in their houses to grant them access to electricity. Figure 9.2 gives the general overview of the data flow on the entire system architecture.

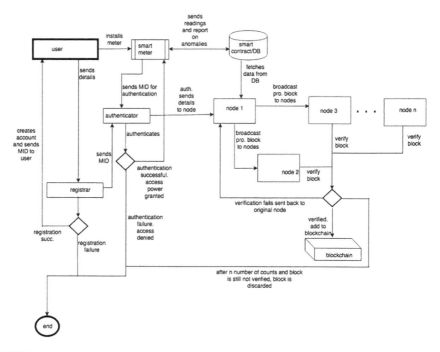

FIGURE 9.2
Data flow on the entire system.

Users register by sending their credentials (name, house number, location, telephone number) through the user layer. The registrar at the registration and authentication center first receives the information. The registrar creates an account by using the user's credentials and generating a unique ID for the client. The registrar further sends the information to the authenticator. Finally, the registrar acknowledges the user by sending the unique ID to the client.

In a situation where the registrar is not available, the user's information is received by a processing and consensus node and creates the account as stated earlier. Authenticator keeps a record of the information and sends a copy to the data processing center. This is received by a processing and consensus node, and the time the information is obtained is recorded. The node programs a smart meter using the unique ID together with the information of the user that is later installed in the house of the user. The unique ID becomes the MID of the client.

The authenticator forwards the smart meter to the house of the user to be installed. When the installation is completed and connection has been established, the smart contract is triggered and a notification that carries the MID is sent to the authenticator for verification. After successful verification by the authenticator, the smart meter is credited with an initial

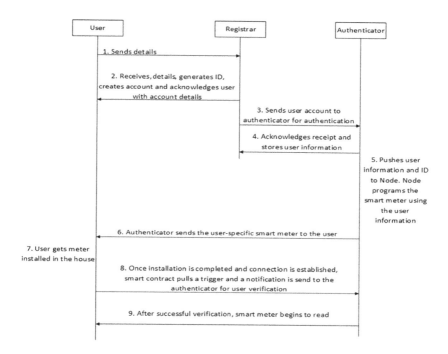

FIGURE 9.3
Step by step registration process.

amount of power of 50 W, and the smart meter starts reading and recording data on a real-time basis. If the authenticator fails, the notification is sent to the processing and consensus nodes for the user to be verified as stated earlier. Figure 9.3 gives the step by step process in user registration and authentication.

The information of the user is handed to a particular node on the grid network. The node adds the amount of power to the user information and processes them into a block. A node's successful construction of a block is broadcast to the rest of the network. This is done to ensure that the block is legitimate. A valid block has been confirmed and approved by the vast majority of processing nodes. When a block is confirmed, it is put to the main sovereign blockchain; otherwise, it is sent back to the creator for review. An invalid block will be sent to the originator until the correction expected is rectified. After some number of counts, if the block is still not accepted, the block is dropped. On our system, after five successive rounds, if the block is still not taken, the processing and consensus nodes go into an agreement and drop the block. The block is given a key, and the key is stored in the smart contract database. Copies of the information are sent to the data center for storage. This information is used for forecasting and research purposes.

Smart contracts on the system send reports to the parties involved if there is a breach of contract. For example, when a user tampers with the smart meter, the smart contract is triggered, and a report is sent to the grid network indicating that the smart meter with this particular ID has been tampered with. The above-mentioned key is used to save this report in the smart contract database. The processing and consensus nodes retrieve the reports from the database and process them into blocks. These blocks are connected to the main sovereign blockchain as a side block when they have been validated and accepted by the majority of processing and consensus nodes. Later on, the side blocks will merge to form the main sovereign blockchain. The smart contract keeps on reporting the various anomalies on the system, and they processed into blocks as side blocks.

9.5 Smart Contract Design

Smart contracts are finite state machines that carry out predefined instructions when predefined conditions are satisfied or specified actions are taken. Smart contracts are used to report on the state of data on the smart meter as well as violations that occur on both the smart meter and the smart grid network. Because the entire data transferred on the system is monitored in a secure environment, this improves the monitoring environment's reliability for both consumers and energy providers. The algorithms are then divided into sections based on the functions they perform, which are described next. Data is first saved in a smart contract database based on the ID that comes with the data before being sent to the smart grid network. Following that, the data is processed, keyed, and broadcast into a private blockchain network. The database of smart contracts is dispersed over the network.

Violations of user data that occur on the smart grid network are processed, keyed, and broadcast into a sovereign blockchain network, and the smart contract alerts the user via the smart meter. On the smart grid network, we write the sets of rules that are applied to the smart meter and the user. When these criteria are broken, the smart contract is triggered to send reports to the smart grid network. They also leave users with alarm notifications on the smart meter's screen. This gives people a sense of what happens to their data when it is sent to the smart grid network. This also applies to messages displayed on the screens of smart meters for the benefit of system users.

After successfully installing a smart meter in the user's home, the smart contract generates a report. It contains the status of the installed smart meter that is whether the installation was successful. This report is delivered to the smart contract center on the smart grid network, where it is placed in the smart contract database using the report's ID. When maliciously modified user data is sent to the smart grid network, the smart contract sends a report to the smart contract center, which is then saved in the database. The smart contract also notifies the user by displaying a message on the smart meter's screen. The smart contract sends a report to the smart contract center and warns the user by showing a message on the smart meter screen when a user's power reaches the minimal value.

When a user's power is depleted, the smart contract is activated, and the electrical power is turned off. It sends a report to the smart contract center stating that the power unit in the residence has been completed. When a smart meter is tampered with, the smart contract sends a report to the smart contract center, which is then put in the database. In this instance, the smart contract instantly turns off the power to the house, the smart meter of which has been tampered with. The processing nodes break down the information in the smart contract database into blocks. As side blocks, these blocks are added to the main blockchain. On the blockchain, the side blocks are attached to their corresponding parent blocks.

9.6 Blockchain Design (Secureness of the System)

This section explains the process taken to design our sovereign blockchain on the smart grid network. The sovereign blockchain consists of two different blocks that are the parent blocks and the side blocks. The first part talks about the parent blocks, and the second part talks about the side blocks.

9.6.1 Parent Block Structure

A block in a sovereign blockchain has a format that distinguishes it from all other blocks. The format is followed by the block size, which specifies the overall block size. The block header is the next structure after the block size. As with Bitcoin headers, the block header is hashed with sha256(sha256()). When it comes to maintaining immutability in the sovereign blockchain network, the block header is extremely important. If an attacker intends to falsify a block's record, he or she must change all block headers starting with the genesis block. This contributes to a greater level of network security by assuring that completing this operation is impossible. The parent block structure of our blockchain network is depicted in Figure 9.4.

A block mismatch will inform the system of a suspicious event, which will activate data forensics in the case of malicious activity. The block header contains the data version of a block, which specifies the validation rules for a certain data type. The attributes and type of data being accessed are specified in the data version of a block. The function is to make it difficult to edit a

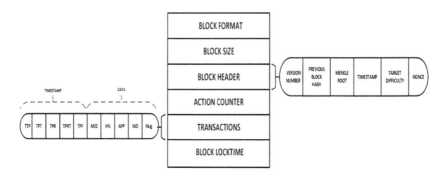

FIGURE 9.4
Parent block structure.

prior block header without changing the preceding block header, and the block's header contains the hash of the previous block, a sha256(sha256()) hash. By ensuring that none of the blocks in the blockchain network may be modified without modifying the header, the Merkle root hash is included in the header. The hashes of all the events in the blockchain network are appended to the current block to achieve this. This yields a sha256(sha256()) result. A timestamp is also included in the header, which shows when the block was produced. In the header, there is a target difficulty that illustrates how the processing and consensus nodes achieve processing. This value is unique to the system and is designed to make processing difficult for malicious nodes while being efficient and solvable for certified consensus nodes. Finally, the header contains a nonce, which is an arbitrary number generated by the processing and consensus nodes to modify the header hash and produce a hash that is less difficult than the intended difficulty.

The action counter in the block keeps track of the total number of violations that have been applied on the accessible data throughout the block. The transactions, which have been divided into two parts: timestamp and data, are the next component. TOC, TTPP, TTPT, TOR, and TTPR make up the timestamp. MID, HN, APP, NID, and Nsig make up the data. Finally, a structure named blocklocktime defines the complete block. This is a timestamp that marks the last transaction entry and block closure. The block is ready to be broadcast onto the sovereign blockchain network once the prerequisites for this field are met. The blocklocktime is the time when a block is accepted into the sovereign blockchain.

9.6.2 Side Block Structure

A side block is formed of a format, and this format is derived by attaching a piece of the main blocks ID to an ID generated by consensus nodes to the side block. The block format is followed by the block size, which is the overall size of the block. The side block has headers, which is made up of six entities. These

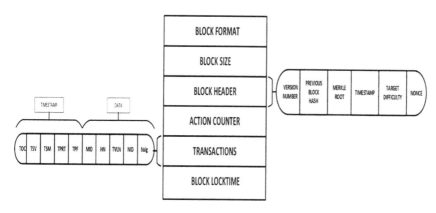

FIGURE 9.5
Side block structure.

elements are the version number, and this uniquely identifies the reports used to build the side block, preceding side block hash, Merkle root of all side blocks for a certain parent block, timestamp, target difficulty and a nonce. These entities have the same features as their parent block but relate to the side blocks. Figure 9.5 depicts the side block structure of our blockchain network. The side block also contains an action counter and this is for documenting the violations on the smart meter, manipulations of data on the smart grid network and the state of the smart meter at any point in time. The next component is the transaction counter, which is made up of the timestamp and the data. The timestamp is made up TSV, TSM TPRT, and TPF. The data is made up of MID, HN, TVLN, NID, and Nsig. The block is then time-locked and broadcast to the blockchain by adding it to the parent block.

9.7 Conclusion

In concluding this chapter, a discussion on the system design established to achieve a tamper-proof and transparent structure was achieved. The system structure created to aid in achieving the overall goal of the project was separated into four areas. The system description was the first section, and it explained how the various components of the system work together to achieve the overall purpose of the project. The data flow across the system was the subject of the second segment. The smart contracts were the third section. This part went over the smart contracts we used in our work to report various system irregularities. The last section was on the sovereign blockchain design. We explained how the blockchain employed in our work was designed.

References

[1] D. Newbery, G. Strbac, and I. Viehoff, "The benefits of integrating European electricity markets," Energy Policy, vol. 94, pp. 253–263, 2016.

[2] E. Brynjolfsson, P. Hofmann, and J. Jordan, "Cloud computing and electricity," Commun. ACM, vol. 53, no. 5, p. 32, 2010.

[3] A. Andrae and T. Edler, "On global electricity usage of communication technology: trends to 2030," Challenges, vol. 6, no. 1, pp. 117–157, 2015.

[4] S. D. Manshadi, M. E. Khodayar, K. Abdelghany, and H. Üster, "Wireless charging of electric vehicles in electricity and transportation networks," IEEE Transactions on Smart Grid, vol. 9, no. 5, pp. 4503–4512, 2017.

[5] P. Sadorsky, "Information communication technology and electricity consumption in emerging economies," Energy Policy, vol. 48, pp. 130–136, 2012.

[6] M. Salahuddin and K. Alam, "Information and Communication Technology, electricity consumption and economic growth in OECD countries: A panel data analysis," Int. J. Electr. Power Energy Syst., vol. 76, pp. 185–193, 2016.

[7] U.S. Department of Energy, "The smart grid: An introduction," Communication, vol. 99, p. 48, 2010.

[8] J. Liu, X. Li, X. Chen, Y. Zhen, and L. Zeng, "Applications of Internet of Things on Smart Grid in China," in 2011 13th Int. Conf. Adv. Commun. Technol., pp. 13–17, 2011.

[9] E. Spanò, L. Niccolini, S. Di Pascoli, and G. Iannaccone, "Last-meter smart grid embedded in an internet-of-things platform," IEEE Trans. Smart Grid, vol. 6, no. 1, pp. 468–476, 2015.

[10] E. McKenna, I. Richardson, and M. Thomson, "Smart meter data: Balancing consumer privacy concerns with legitimate applications," Energy Policy, vol. 41, pp. 807–814, 2012.

[11] F. Li et al., "Smart transmission grid: Vision and framework," IEEE Trans. Smart Grid, vol. 1, no. 2, pp. 168–177, 2010.

[12] W. Wang, Y. Xu, and M. Khanna, "A survey on the communication architectures in smart grid," Comput. Netw., vol. 55, no. 15, pp. 3604–3629, 2011.

[13] Y. Kabalci, "A survey on smart metering and smart grid communication," Renew. Sustain. Energy Rev., vol. 57, pp. 302–318, 2016.

[14] X. Hao et al., "Smart Meter Deployment Optimization for Efficient Electrical Appliance State Monitoring," [C] in 2012 IEEE 3rd International Conference on Smart Grid Communications, SmartGridComm 2012, pp. 25–30, 2012.

[15] C. I. Fan, S. Y. Huang, and W. Artan, "Design and implementation of privacy preserving billing protocol for smart grid," J. Supercomput., vol. 66, no. 2, pp. 841–862, 2013.

[16] V. C. Güngör, D. Sahin, T. Kocak, S. Ergüt, C. Buccella, C. Cecati, and G. P. Hancke, "Smart grid technologies: Communication technologies and standards," IEEE Trans. Ind. Informatics, vol. 7, no. 4, pp. 529–539, 2011.

[17] S. Rusitschka, K. Eger, and C. Gerdes, "Smart Grid Data Cloud: A Model for Utilizing Cloud Computing in the Smart Grid Domain," in 2010 First IEEE International Conference on Smart Grid Communications, pp. 483–488, 2010.

[18] A. R. Metke and R. L. Ekl, "Security technology for smart grid networks," Smart Grid, IEEE Trans., vol. 1, no. 1, pp. 99–107, 2010.

[19] M. Mylrea and S. N. G. Gourisetti, "Blockchain for smart grid resilience: Exchanging distributed energy at speed, scale and security," in 2017 Resilience Week, pp. 18–23, 2017.

[20] Q. Xia, E. B. Sifah, K. Huang, R. Chen, X. Du, and J. Gao, "Secure Payment Routing Protocol for Economic Systems Based on Blockchain," in 2018 International Conference on Computing, Networking and Communications (ICNC), IEEE, pp. 177–181, March 2018.

10

Tokenization of Energy Systems

10.1 Introduction

A smart city comprises several entities with multifaced properties that interact with each other amid complex actions to produce a set of defined results. This is true of energy systems where several components come together in major operations, ensuring the effective generation of energy across a smart city. However, one thing remains true, in the natural smart city, several energy sources exist, each affecting a single environment. These sources of energy produce several types of energy [1] that are used by various devices for various use cases and activities. This type of energy produced is of two major forms. They are either generated from renewable energy (RE) sources or from non-renewable energy (NRE) sources [2]. Thus, devices in a natural smart city are intertwined between these two energy sources. These devices and agents include wind turbines, solar panels, vehicles, virtual power plants, distributed service operators, and several distributed energy resources that exist side by side. While many devices currently make use of NRE sources, NRE systems have matured and have become inefficient and costly. In contrast, RE generation has also become cheaper. However, the challenge is that since these different energy sources exist in a single environment, their production affects the environment and the participants within the smart city. However, it is difficult to control the behavior of members since no clear rule or policy can simultaneously regulate behavior within a smart city and energy system. Thus, most solutions have depended on external third parties to regulate the energy systems and the behavior of participants within these smart cities. It is also a great challenge to identify the various devices using the diverse source of energy in very complex smart cities where several devices operate simultaneously and independently of each other. Thus, an attempt to control thus multiple devices within a single smart city with multiple participants participating in an energy network with various sources of energy type is extremely unrealistic to achieve. Multiple attempts have been made to resolve this, including the separation of various smart cities into single units that can be regulated, but that is very hard to achieve.

DOI: 10.1201/9781003289418-10

Decarbonization of various non-traditional energy devices and devices that make use of NRE like fossil fuels still remains a challenge since there is not a decentralized [3], effective monitoring system to be able to control and incentivize parties to reduce the rate of CO_2 emissions.

Again, identifying various entities sometimes producing dual-energy types and having multiple devices is very hard to achieve in an extremely large smart city. There needs to be an effective means of identifying individuals within a smart city, observing their various actions in a very transparent manner, and at the same time be able to regulate their behavior to that the benefits the entire smart city in a very transparent manner-something the blockchain provides.

Again most of the incentivization schemes needed to ensure that individuals and energy systems within a very complex smart city that generate multiple energy sources across a smart city are very hard to achieve. Therefore, there is a need for an effective incentivization mechanism to be created that can be enforced without the need for a trusted third party. In addition, the incentivization must be able to ensure behaviors that ensure energy conservation, decarbonization and effective distribution of energy generated.

We resolve these challenges by proposing a blockchain-based incentivization and exchange system making use of a blockchain-based tokenization scheme and a decentralized Energy Automated Market Makers [4] in NRE/RE smart cities. This is to regulate both NRE- and RE-based environments. Tokens affect behavior and provide a means of reinforcement among participants on a network. Tokens provide a means of reinforcement across an entire smart city where various components must come together to work together effectively. Participants within these environments can interact and share energy resources in an effectively controlled manner without the need for a trusted third party. We also developed a novel incentivization scheme [5] using a blockchain-based tokenization system where participants are rewarded and punished based on pre-set rules. We ensure required behavior by using blockchain-based smart contracts as an efficient means of setting effective protocols. Since the blockchain with its set of smart contracts provide a means by which an effective tokenization system [6] can be developed, it is easier to incentivize NRE users to utilize RE sources and be able to interact with RE device without a high barrier of exchange and dependence of external third parties. Multiple parties on the blockchain can interact with each other and be able to exchange energy of various types. We also control energy systems using the same tokenized system to develop a complex tokenized mechanism making use of reputation mechanisms, resource generation, and access control. We also develop a unique Energy Market using Automated Market Makers, where the various energy sources generated within the smart city are traded as tokens and converted across the various energy systems, thus increasing or decreasing the number of tokens of each type within the smart city.

By attaching the real amount of energy produced from the various sources that are utilized within the smart city with real energy supply generated by NRE and RE systems that can be exchanged [3], converted and exchanged over an Automated blockchain-based Energy Market, we develop a unique architecture where participants within the smart city, no matter the type of energy being utilized, can interact with each other. They can be rewarded for their effective behavior while at the same time keeping the amount of energy produced and utilized within the smart city balanced and controlled in a trusted and effectively secure manner over the blockchain. In addition, we ensure that the Energy Market can interface with other Energy Market under a group of specialized contracts over an arbitrage mechanism. This provides an efficient, efficient, and highly regulated transfer of energy of various types across multiple smart cities while keeping the same network balanced. In this work, we develop our unique concept incorporating various unique innovations.

1. We create a unique energy-based smart city that brings together participants using NRE-/RE-based energy resources to interact with each other in a blockchain-based environment under specific rules set by a set of smart contracts that enable them to collaborate and share energy effectively.

2. We design architecture that divides energy smart cities into a specialized component. We separate each component into a module based on the specific function and entities existing within the module. We separate this into Energy Supply Modules, Energy Storage Modules, RE Residence Modules, NRE Non-Residence Modules, and Energy Markets.

3. We develop a novel Automated Energy Market Makers (*AEMM*), where the energy of various forms is traded, converted, exchanged, and regulated between multiple smart cities. Using a set of specialized smart contracts, we can regulate energy transfers between two smart cities exchanging energy over a blockchain-based Energy Market.

4. We develop an exchange mechanism between NRE and RE where both energy types can be exchanged and converted from one to another without destabilizing the energy smart city. Furthermore, we set token units to match real-time energy units odious energy types stored within the energy smart city, facilitating a means for the various energy generated to be managed and effectively used on our blockchain network.

5. We develop a unique tokenization mechanism that enables various energy types to be funded, generated, and utilized by participants within the smart city. We also utilize the token mechanism developed on the blockchain to be able to able to regulate and incentivize the energy-based smart city.

6. We develop a set of rules for which good behavior within the energy smart city can be utilized using blockchain smart contracts by setting rules that ensure decarbonization of NRE energy produced and provide a means for the energy produced within the smart city to be conserved.

10.1.1 NRE/RE Tokenized Blockchain-Based Energy Contracting

For each energy smart city, there is the need for contracts among multiple devices to ensure control of the energy flowing in between this system. Several policies and instructions must be set to regulate and control multiple activities of multiple devices. Thus, contracting [7] within energy systems and amid multiple participants must be made in a transparent, decentralized manner without the need for a third party. Discrete choice probabilities will be used to make this contracting decision. If the expected profits (net of transaction cost) from contracting are larger than the expected gains (net of transaction cost) from structuring transactions in another way, parties to transactions will choose to contract.

This concept is expressed as follows: $G^* = G^c$, if $V^c = V^a$ and

$$= G^a, \text{ if } V^c < V^a \tag{10.1}$$

where G^c is the primary contracting value. G^a is the alternative contracting value, V^c and V^a are the transactor's presumption in each contract creation represents the best contracting value chosen. We express our argument as

$$V^c = V^c(X, e_c) \tag{10.2}$$

$$V^a = a^X + e_a \tag{10.3}$$

where X is a vector of observable attributes, which affects the game achieved from pre-set arrangements a^x and e_a. On the blockchain, smart contracts provide a means by which contracting can be done effectively and for the right incentive mechanism to be made and payoff achieved in a trusted and verifiable manner.

10.1.2 Tokenized Systems

Tokenized systems are needed for large smart cities on a large scale [8]. However, few works have explored the effective use of a tokenized system for blockchain-based energy systems. Since energy smart cities consist of several homes and industries, each making a tremendous amount of energy usage, a blockchain-based tokenized approach provides a means to make effective measurements and control all entities within the smart city. With identifiers provided by the blockchain, we can obtain feedback information, such as data collected at the time of fuel delivery, the amount of fuel used in comparison to the same period the previous year, and the dollars saved as a result of the lower utilization. With this, tokens can be used to regulate how much energy must be consumed within the next coming period. Blockchain-based tokens [9] provide a means to control the entire smart city to ensure efficiency. Natural enforcers and other energy-control third parties can be substituted for effective third-party systems. The blockchain serves as an autonomous system that can alter the behavior of individuals within an ecosystem. This provides an innate means of attaining desirable performance without interference from any external system. Several mechanisms can be taken with this approach, which our work here takes a look at. Variables are set with smart contracts and ensure management and accountability at all entities. We can establish cooperative behavior among entities within the smart city. The blockchain provides a means for immediate reinforcement, and smart contracts provide a self-verifying mechanism for delaying and hastening consequences. Rewards can be set by building target behaviors and rewards for which tokens can be built and attributes can be set. An incentivization mechanism can be easily developed since blockchain-based tokens are backed by smart contracts, which initiate them. Underlying the blockchain with an effective tokenized system that controls several energy sources ensures energy-saving behavior and effectively regulates energy produced. In-built directives such as feedback variations, prompting techniques and reward systems such as reputation-based tokens, hierarchical classification, and token-based payments can ensure energy conservation. To be able to implement a token-controlled energy system, it is important to change the settings completely to provide a more adaptive framework.

10.1.3 Automated Market Makers

For Energy Market to make use of tokens for incentivization of participants and controlling the energy network, it is important to ensure, the effective tokenization of all energy generated and regulate balance in the energy being exchanged in order not to destabilize the entire network and cause loss of energy as well as constant fluctuation. In our work, we make use of a Constant Function Market Maker *CFMM* due [10] to the number of energy sources for which we tokenize as i and other metrics that must be tokenized within our system. We define our *CFMM* by a trading function, $\rho: R_+^n * R_+^n * R_+^n \rightarrow R$ and its reserves as $R \varepsilon R_+^n$. We balance all token reserve over the blockchain-based Automated Market Maker via a Constant Product Market (*CPM*) Model implemented using smart contracts to ensure balancing of token reserves as well as physical energy stored. We also define our *CPM* value with an underlying tuple (t, R_β, R_α). An entity exchanging Δ_β at time t will receive Δ_α, which satisfies transactions $(R_\alpha - \Delta)(R_\beta + \Delta_\beta) = k$ where $R_\beta, R_\alpha > 0$, $k = R_\alpha R_\beta$ for all reserves having $R_\alpha \geq 0$ and $R_\beta > 0$ constant product $k = R_\alpha R_\beta$ and percentage fee $(1-\gamma)$. After each transaction within the Energy Market, the market is updated as $R_\alpha \rightarrow R_\alpha - \Delta_\alpha$, $R_\beta \rightarrow R_\beta - \Delta_\beta$, and $k \rightarrow (R_\alpha - \Delta_\alpha)(R_\beta - \Delta_\alpha)$, $(R_\beta + \Delta_\beta)$ and $k \rightarrow (R_\alpha - \Delta_\alpha)(R_\beta + \Delta_\beta)$. In order to ensure the total energy balance and prevent a non-positive tokenized energy reserves, we require that three $R_\alpha R_\beta > 0$. This also helps to prevent the case of infinite cost. For each *CPM* value [4], with a zero fee, that is, exchange of entities within each Energy Market M, must change the reserves $(R_\alpha + R_\beta)$ such that the product of the Energy Market token reserves remains equal to k as a constant.

10.2 Proposed Solution

This part introduces and discusses the steps involved in our proposed solution. Figures 10.1 and 10.2 depict the graphical view of the proposed solution.

10.2.1 Registration

Participants register on the blockchain network. We collect information about entity registering, such as their names, blockchain-based addresses and other relevant information. Other information includes their roles. We also collect other information such as the entity they represent, such as residence id and location. Other energy-based information, including the amount of energy *amt* produced, energy types utilized, amount produced for each, daily emission per day for particular *ResidenceID* or *IndustryID*, is then recorded on the blockchain. We classify all entities as RE/NRE entities based on the paramount device associated with that entity. We generate a Registration ID, *RegistrationID* for entity to be registered. We issue a classification receipt to all entities based on the kind of energy resources used by the particular entity or the corresponding device. Suppose the number of devices associated with that particular *EntityID* on the blockchain network exceeds 50%. In that case, he or she is classified according to the highest kind of classification type his ID is linked to. We also classify the various energy devices based on the type of energy that these devices will utilize. We provide a particular ID *EnergyID* for each energy type and assign it to the particular device ID. Their information is then added to the blockchain network. A set of procedural protocols are activated for the particular entity. This initiates a set of smart contracts the particular entity *EntityID* can be able to interact with. Our blockchain syncs all this information and balances the Energy Market and the entire network. Tokens are generated to the various energy pools based on the amount of energy needed to balance the entire network. This corresponds to the particular token type, which is stored in our blockchain-based vault.

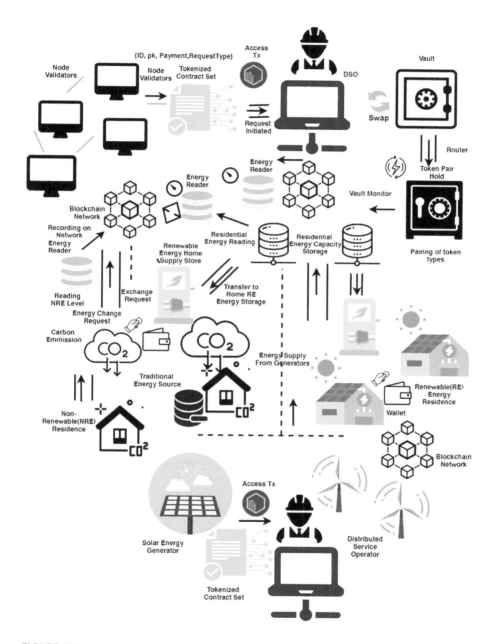

FIGURE 10.1
A tokenized blockchain-based smart city with non-renewable and renewable entities and devices.

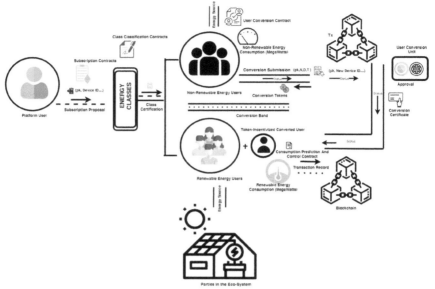

FIGURE 10.2
Tokenized-registration mechanism for classified participants within smart city.

10.2.2 Setup

Here we develop a framework on which all other operations can be made in the blockchain-based ecosystem. This works in a way to make the registration process easier on the blockchain ecosystem by registering all entities. We initiate a set of smart contracts on our blockchain known as *SetupContracts*. Here we designate the types of entities. From device, residence, industry, individuals, these are designated a particular id, and the entity type *EntityType* is set up for each of these entities with the corresponding name. Individuals on the blockchain are also classified based on their particular roles. We create a set of roles as *RoleSet* such where we group entities into consumers, token pool developers, industry managers, and homeowners. For the particular location of entities, *loc*, we calculate the energy needed to power devices within the entire location. We obtain this information from the *DSOs*, which are broadcasted across all nodes. The amount of energy capacity required *EnergyCapacity* for a particular location is then logged onto the blockchain. For a particular entity to be entered into the blockchain, it must meet requirement *RegReq* set up by the *RequirementContract* to participate on the blockchain network. Suppose new actor entities like smart meters or PV panels want to enter the network. In that case, the *RegistrationContract* checks against the requirements set for the device using the *RequirementContract*. If the applying device is accepted, the necessary information of the particular entity from role, name, and all other information is logged into the blockchain.

The estimated energy required to power all entities *OverallEnergyCapacity* is calculated by the *DSOs* and broadcasted to the entire blockchain network. To provide an incentivization mechanism, we set up a tokenized system over an Energy Market. These tokens are generated in a Token Vault, which houses the reserves of various energy types within each *Energy Market*. The tokens are set to each of the token types listed within the *TokenContract* on the blockchain. We generate a set of id *EnergyTypeID* to identify each of the energy types and pass the parameter *TokenGenInfo* to the Energy Market to be generated. We also have a set of identifiers for each Energy Market on the blockchain network. We submit this to our blockchain network as *SetupTransactionInfo*.

10.2.3 Proposed Smart City Module

We set the initial energy generation i at time t. We read all ongoing transformations from our blockchain network. We observe the transmission of energy flows across our blockchain network. We set up the architecture of all entities on our network. We set up our architecture into five modules: The Energy Market, NRE Residence Module, RE Residence Module, Energy Storage Module, and Energy Source Module. Figure 10.3 shows the overview of the proposed token-incentivized blockchain network.

1. *Energy Market:* This serves as the main powerhouse for the blockchain-based ecosystem [11]. This is made up of a set of contracts and blockchain storage that provides a trading interface for participants on the network. It is composed of the Energy Type Token Vault, an Energy Trading Center, and a set of smart contracts that manage the interactions between them known as Trading Center. It reads information from the Energy Source Module to provide a set of demand and supply by matching the amount of energy produced across the network to that of the Energy Storage Module and ensuring that consumption and supply of energy to the network are well managed.

2. *Non-Renewable Energy Residence Module*: This refers to the non-renewable components of the ecosystem. This includes residence and industries that make use of NRE sources such as fossil fuel. This is also referred to as the set of devices and equipment that monitors and controls the emission of NRE produced within the energy ecosystem. Entities within this module are incentivized to reduce energy by being allowed to participate in the Energy Market. Here they are rewarded as they reduce the emission of non-renewable produced. They are also able to obtain RE supply from the *Renewable Energy Residence Module*. Thus, they can leverage the energy supply from Energy Supply, ensuring a reduction in NRE sources from the ecosystem. Furthermore, they regulate the energy output within the

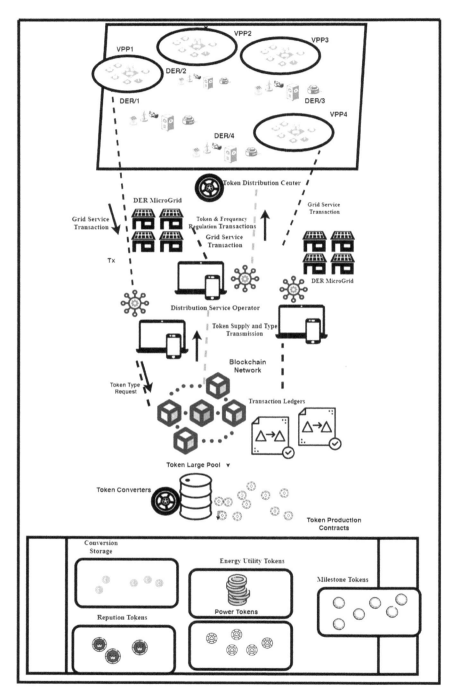

FIGURE 10.3
Overview of our token-incentivized blockchain network.

ecosystem by reducing energy usage during specified periods, for instance, nighttime. They also reduce the processing usage when not in use. All these behavior within the ecosystem are rewarded by the *NREBehavioralContracts*. For each action on the blockchain network, ensuring a behavior *p1*, a higher occurring behavior reinforces a lower probability behavior on the network. Hence, the network can control sub-behavioral patterns among parties.

3. *Renewable Energy Residence Module*: This contains entities that make use of RE in the daily operations of the ecosystem. The Energy Source Module powers them. They include industries and residences within the locality *loc*. They also include RE devices. This module measures transactions and energy supplies that take place among entities that make use of RE in their operations. They provide a means of monitoring all activities involving RE devices across the ecosystem. Devices and transactions within all residences and industries making use of renewable energies from different sources are registered in the *Renewable Energy Residence Module*. We monitor transactions within them. Prosumers within this module also transact with the *Non-Renewable Energy Residence Module* supplying and exchanging energy across the *Energy Market* for a reward. This also includes the *DSOs* that ensure the effective distribution of renewable energies across the blockchain ecosystem. The prosumers within the *Renewable Energy Residence Module* contribute to the energy being supplied to the entire network. They increase or decrease the energy supply based on the amount being demanded *AmtDemanded*.

4. *Energy Supply Module*: This involves the set of devices responsible for generating energy and power for the entire ecosystem. This includes both RE and NRE sources. These include thermal plants, wind and solar devices, and several other devices that generate traditional power sources. The energy generated is transmitted into the ecosystem [12]. They are read by the *DSOs* and then allocated to the *Energy Storage Module*. From there, they are utilized within the ecosystem.

5. *Energy Storage Module*: These devices within our blockchain-based ecosystem store energy from various energy sources. Through a set of automated contracts, energy coming in from various energy types is monitored. All transactions are observed, and the energy coming from different nodes across the network is read and recorded on the blockchain. The energy type is split by identifying the energy sources coming into the blockchain ecosystem [13]. The Energy Storage Module then matches the information it has read at a specific time to that of the *Energy Market*. The tokens of the various energy types are then adjusted accordingly based on the readings from the *Energy Storage Module*. Upon exchange transaction from NRE sources, a set of processes are initiated, and the needed energy supplied is made to the particular entity within the *Non-Renewable Residence Module* without the need for a third party.

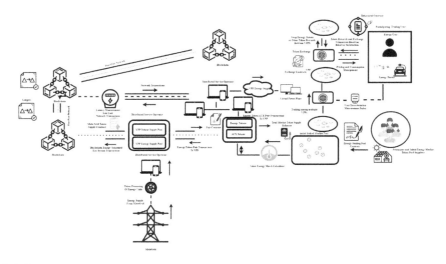

FIGURE 10.4
Inner-architecture of our blockchain-based automated Energy Market maker.

10.2.4 Blockchain-Based Monitoring of NRE/RE Ecosystem

We set all entities up within our blockchain network based on the modules left. We monitor all activities within the smart city. We then identify the various entities on the network-based energy production type *EntityType* as shown in Figure 10.4. If the energy type being generated by each device, residence and industry with a particular id *EntityID* is found located on the blockchain that falls under a category *NRESource*, we add the particular entity to the *NREEntityModule*. We monitor all entities on the network. The energy within this network is stored and then utilized across the network. Each residence is registered on our blockchain network. The location, house type, types of energy produced within each household are registered on the network. This provides a unique identifier to identify every particular residence within the smart city, which serves as a node on the blockchain network. For each house located on the blockchain, we register each residence. Each residence and industry within the residence submits its location, id, name, image, and other essential details to the blockchain.

10.2.5 Tracking of Energy Generated

We track the type of appliance used, the amount of energy being generated, the location, the expected volume of energy expected within the particular geographic location, the carbon density as well as the price per token based on the readings from the energy reader and adjustment from the *DecarbonizationContract*.

The energy readers monitor each amount of energy produced within a particular residence [9]. In the feedback-incentive condition, households get weekly energy consumption information and fixed rewards based on the percentage decrease in energy (gas or electricity) compared to baseline. In addition, additional awards are awarded for the most significant reductions made in a given week. We set up the connection point on our blockchain-based smart city. We set up the connection point for the modules. We set up the metering infrastructure for all the residents within our smart city, which provides all the necessary information. From this, we are able to see the performance of all the household appliances and devices within a particular industry. The energy reader reads every piece of information, and transactions are controlled by the *EnergyReadingContracts*. The *EnergyReadingContracts* transmits the energy reading from the NRE Module to the Energy Market. All the information from the price, *P*, total demand, *TD*, total available generation, *TAG*, carbon limit set, *CarbLimit*, and present amount of Carbon Emission, *CarbEmmission*, as well as other network conditions are transmitted across other parts of the network. The price and demand information provides a means for providing consistent real-time demand for RE being traded on the blockchain.

10.2.6 Decarbonization within NRE Ecosystems

For each of these households, exceeding the carbon limit provides a set of penalties that is meted out if they exceed the carbon limit *CarbLimit*. However, if they can reduce their carbon emission within the household, a token reward is provided to them and transferred to their blockchain wallet. Residents are given a full report of how much energy has been consumed, how much has been supplied, and the amount of NRE reduced in each household. A set of ratings is issued for the users on the blockchain network. Several NRE limits are set based on the NRE type. For instance, several grades of carbon limits are set based on the particular entity if it is carbon. That is whether a residence within the smart city or an industry. The *Non-Renewable Energy Contracts* monitor the level of carbon emitted at each time *t*. Upon exceeding the limit, a penalty is enacted by the *Non-Renewable Energy Contract* based on the level of carbon limit reached. The blockchain notifies a warning notification to the particular entity.

The smart contracts provide a set of options for the particular entity. He or she is either asked to reduce the emission by cutting down on the amount of traditional energy produced to avoid other penalties, or a decarbonization energy exchange is requested. For each carbon limit that exceeded a certain amount, penalty fee is taken from his blockchain amount. This helps to control the rate of energy production, and the amount of energy produces on the blockchain network. This is controlled by the *DecarbonizationContract*.

Once the *DecarbonizationContract* is activated, an instruction is submitted to the *ExchangeContract* [14]. A certain amount of fee is taken from the consumer's blockchain balance. This is visible to all participants on the network.

10.3 Proposed Decarbonization Functional Procedure

The blockchain receives a notification from a particular entity making transactions on the blockchain network. A set of requests is submitted to Entity A by the *DecarbonizationContract*. Entity A is either asked to reduce the carbon limit by reducing the various parameter M that makes the carbon limit *CarbLimit* run in excess. A set of time *ReductionTime* reduce is provided within which Entity A is required to reduce the carbon through the adjustment of parameter M being read by the energy reader. If the parameter M cannot be achieved, the *DecarbonizationContract* issues an exchange protocol for an energy exchange to the energy *ExchangeContract*. The energy *ExchangeContract* checks if Entity A meets requirement *ExchReq* needed to participate in the energy exchange from the set of information received from the *DecarbonizationContract*. If he or she does not meet the necessary requirement, the exchange protocol is canceled. The reminder for the carbon amount left $t_{carbreduced}$ is calculated. If the parameter M is still not reduced. A penalty of a particular fee f is issued to Entity A on the blockchain network. Subsequent fees $f + 1$ are issued to Entity A until the *CarbLimit* is reduced to a particular level L. If Entity A meets the requirement *ExchReq*, the subsets of contracts within the *ExchangeContract* are invoked. An *accessID* is issued to Entity A. The blockchain measures the energy level of Entity A. Entity A submits a request that contains the amount of energy in the system *amt*, the id of the particular entity, the access id, and the amount of energy to buy other relevant information.

10.3.1 Design of Proposed NRE/RE Decarbonization Contract

The *DecarbonizationContract* ensures the effective control of the carbon emission from the *NRE Module*. For the residence and industries that reduce emission at a certain threshold *Threshold*, our blockchain-based smart city provides a set of interest reward *interestreward* that is transmitted to the wallet address. A number of tokens are received for compliance with the requirement *DecarbReq* set by the *DecarbonizationContract* to ensure the effective control of carbon emitted. For carbonized residence for each specified location, *loc*, an index is set for that particular residence and industry as *EntitityID*. A different level of limit is set for residence or industry based on a role as *EntityCarbonLevel*. Various thresholds *Threshold* exist for various roles such as residence or industry. Upon exceeding the decarbonization level, a notification is emitted by the blockchain network and is received by the *EntityID* at a particular location *loc*. Upon reception, the *DecarbonizationContract* is initiated for that particular entity on the network. The *EntityID* responds first by reducing the amount of NRE being used by *EntityID*. If a certain level of reduction *red* is achieved, a reward token of an amount *rewardtoken* is sent to *EntityID* to the Energy Market on the blockchain. This is sent to the address of the *EntityID* and signed by his or her wallet.

However, if after time *DecarbTime*, a reduction is not achieved, a set of protocols are initiated toward full decarbonization by the DecarbonizationContract. Since limit and decarbonization time are exceeded, a transfer *pen* is transferred to the wallet of the Energy Market. This provides more funding to generate more energy for the entire smart city. In order to incentivize the participants to reduce the amount of carbon emitted within our proposed energy smart city, we set up a token-based reputation system that ensures that each entity is rewarded based on the degree of decarbonization behavior practiced within our blockchain-based smart city. Thus, entities receive a Reputation Token of an amount *reputtoken* that is also transmitted from the Energy Market. Each *reputtoken* received matches to a particular grade level *gradelevel*. A set of benefits *benefit* is also assigned based on the particular grade level by our control contract *ControlContract*. For instance, at a particular grade level, the amount of fees *f* needed to transact with our blockchain-based *Energy Market* is reduced to a particular amount *disc*. In addition, participants of certain grade levels receive a set of specialized Access Tokens that enable them to participate in transactions related to that particular tokens in the Energy Market. This is ensured by the contract that regulates transactions within the Energy Market, that is, the *EnergyMarketContract*. Since all participants are connected to the blockchain, monitoring and controlling transactions occurring within each residence and industry are easier. Since the blockchain provides a trust-less and controlled smart city, the decarbonization of entities is easier to manage and control.

The changes in energy within the NRE Module are accounted for by the *Energy Market* [4]. The transaction hash of the transactions made is then stored on the blockchain and seen by all parties on the network. In order to reduce the amount of energy sent within the blockchain, *EntityID* must participate in the *Energy Market* as a *Buyer* in order to exchange an amount *amt* of Non-Renewable of type *EnergyType* in order to avoid being punished after *DecarbTime* exceeds. *EntityID* submits a request to the Exchange Contracts *ExchangeContract* in the *EnergyContract*. The necessary parameters include *EnergyCapacity, AccessID, AmountRequested* and other relevant information. The *ExchangeContract* calculates the current energy prices and current demands on the network and submits this information to the Energy Market. Once the energy trading between is complete, the transaction results in *DecarbTansaction* are sent back to the *Non-Renewable Energy Module*. The information is then logged onto the blockchain. Upon receiving the request

ExchReq, the *ExchangeContract* invokes another set of contracts known as the *TradeContract*. The information is broadcast across the chain. The *accessID* is then transmitted to the blockchain. The transaction hash of each transaction is also traced to be able to know the details coming in from the *Non-Renewable Energy Module*. The *TradeContracts* are then initiated to begin the energy trading process.

10.3.2 Proposed Blockchain-Based Energy Market

The *NRE Module* and *RE Module* all connect to the Energy Market. All entities such as *NRE/RE* residence and industry are identified on the market using their *EntityID*. The entities that would like to provide *Tokenized Energy Pool* funds to the *Energy Market* are set up as *Funders*. However, most entities within our network primarily aim to exchange the energy they have for the various energy types as well as reduce the carbon emission within their network. The Energy Market used in our work is an *Energy Automated Market Maker*, which provides a decentralized manner of ensuring the exchange of energy across all the networks by trading the energy among complex incentivization schemes. A complex tokenization system, a balancing pricing mechanism, and connectors to the various units power the *EnergyMarket* used in our work across our network.

These connectors are made of sensors that provide the accurate reading of transactions across multiple units. The *Automated Energy Market Maker* possess within its structure a *Trading Vault* and the *Energy Token Pairing Center* and a blockchain-based *ExchangeContracts,* which manages and controls the market. They also manage the exchange of tokens received via request from other contracts within the unit modules. They control the *Energy Demand* across the various networks and contain contracts responsible for token generation, fee management and managing all activities on the market with the *DSOs*. Upon initial setup, the several token types are registered on the network based on the available token types that are detected by the energy readers found at the *NRE Module* and *RE Module*. The system adjusts itself through the computations from the *MatchContract* to balance the energy from the *NRE Module* to *RE Module*. It takes the time at which each adjustment *TotalEnergyAdjustment* is made and one set to optimize value *opt*, a notification is sent across the network as time *t*.

All other information is then also logged onto the blockchain, such as the time *t* that the adjustment is reached. The first state on the blockchain at the time the initialing adjustment adj_i is broadcast. This serves as the initial point for the *Automated Energy Market Maker*. The Energy Market makes reference to another *Market B* from which it is able to compare prices transparency and from which participants can share energy. As such, different Energy Markets can trade respective energy amounts across several blockchains. Based on the *PricingContract,* we calculate the amount of energy necessary to collate the entire network and calculate the corresponding price as well *StakePrice*. This serves as the amount needed to be staked by all funders. Using our pricing mechanism, the algorithm calculates the amount of incentive i per each funder based on reputation R, type of user U_{type} as well as the amount *EntityStaked*. From this, we can be able to collate from all funders the needed amount of funds M_{fund} needed to generate the Tokens T of n types to the *AEMM* for control and management. We believe in our work that sets a balance *bal* for all entities and devices on the blockchain to provide more autonomous, decentralized and effective control of all entities on the network. It also provides a good means of monitoring all activities happening on the network by all the parties involved. It also enables all observable parties within the various modules to observe the current energy within the smart city as well the total energy for both the *NRE and RE Module*.

As the various units take such effective decisions in regulating the amount of energy being generated within the smart city as well as the amount, from this, we set up a $Tokenized Energy Pools funding system for the various energy token types on the blockchain. This is because each energy token type supplies a different set of demand D_i to the market at a particular time t. Hence we set up a different *Tokenized Energy Pools* for each of the modules to power them. The *EnergyControlContracts* contains a set of rules that split the total amount of energy generated *TotalAmt* to the various amount of energies based on the Constant Product Function [4]. *Funders* can provide the necessary funding of l to the various pools. The same pricing mechanism is used across the board on the blockchain since it is important for the various set of *Tokenized Energy Pools* set on the blockchain $TEP_i, \ldots, TEP_n$ to match one another on the blockchain. Once the various set of tokenized energy pools $TEP_i, \ldots, TEP_n$, a request is submitted to all entities within the network as *FundReq* with the following price, energy amount, tokenized energy pool id, market id as *Tokenized Energy Pool* transaction *P, Amt, TEPID, Tokenized Energy Pools Type,* and *MarketID, Device*. Other sets of information are added and hashed into the blockchain. This is sent as transaction T on the blockchain network.

The entities respond by submitting their funds *funds* to the particular *Tokenized Energy Pool ID* selected by a particular entity funder *funder$_i$* based on their preference. Once all energy tokenized energy pools $TEP_i, \ldots, Ln$ have been fully set, we initiate a set of *Tokenization Contract* and activate the *Control Contracts* across the various modules. Once detected, the particular *TokenType* is submitted to the *DSOs* for approval of the particular *TokenType*. Each token type is given a particular id *TokenTypeID*. The *TokenTypeID* refers to a particular token of a particular type. Based on the amount of energy being generated in the smart city, a certain amount of tokens are produced. We call this process token type minting. The amount of tokens minted for each token type is recorded on the blockchain as *TokenTypeQuantity*.

10.3.3 Token Generation

Each vault has the capacity to perform a set of unique functions in the management of the energy tokens. We use the constant function product to generate the number of reserves of each energy type. We set the constant function to be the amount of energy provided on the energy smart city for that particular type of energy. For this, a specified number of tokens are minted on the blockchain network. Due to the fluctuations from the reading of the amount of energy on a particular grid, there is constant production and destruction of tokens, which creates a variation in demand and constant change in prices. The wrapped tokens provide a current understanding of the varied changes in energy on the blockchain network. Thus, each energy type has its relative token measurement based on its current price level and quantity on the blockchain. This means that the tokens produced a match to real-time assets and are thus set as wrapped tokens and thus must respond to the changes taking place on the network.

The token created follows a certain defined standard as specified within the *EnergyContractFactory*. Thus, virtualization of all physical assets is made with the tokens using the blockchain to enable effective control and management. To begin to create new tokens upon request, the *TokenTypeID* is submitted to the *EnergyContractFactory* that controls the *EnergyTokenVault*. This is responsible for producing the energy tokens of each particular *TokenTypeID*. The *EnergyContractFactory* checks the fees and sees if the fees required to create the tokens have been satisfied. It generates the tokens based on certain properties such as *amt*, type, bid or not, and whether it is *Fungible* or *NonFungilble*. If the tokens have not been satisfied, it issues up a request to the *DSO* for full payment of the fees. The *DSOs* submit the fees *f* that is needed for the creation of the tokens. Once the fees have been paid, the *EnergyContractFactory* begins to create all the contracts. The Energy *TokenType* is set based on whether it is Fungible or Non-Fungible. These provide a difference in which various *Entities* within the smart city will be accessed by various *Entities* within the smart city. A set of operations are also set based on these criteria.

For instance, if it is Fungible, the list of participants who have access to each token type is also broadcast to the blockchain. A certification ID is issued to all the *DSOs* responsible for paying the fees to produce the tokens on the network. Only the *DSOs* must be able to initiate token creation. A multi-signature system for all *DSO* on the network is set up for approving the creation of the *TokenTypes*. Once the approval is reached by all *DSOs* on the network through cryptographic proof, we initiate the token generation of the various token types. The owners of the token types are set to the secret key using a multi-signature scheme for all multiple *DSOs* on the network. Once a person is able to provide the secret key *p* for the signature multi-signature verification, the *TokenTypes* can be generated after approval from all *DSOs* on the blockchain network. We use a multi-ownership system for the token access in a multi-ownership to provide security and transparency for participants on the blockchain. Using the same multi-signature means of approval, an owner can make a request to make a certain amount of token unexchangeable. If accepted by all parties, a token is restricted from being bid upon in the Energy market.

10.3.4 Energy Funders

The Energy Funders, *Funders* are responsible for the setup process for setting up the entire blockchain-based tokenized energy smart city. They provide the necessary funding needed to generate the amount of energy needed from the *EnergySupplyUnit*. They respond to a request from the Energy Funders *FundReq* with a certain *TokenizedEnergyPoolIDs* and a certain amount of energy *energyamount$_i$* of a particular energy type in order to transact a certain amount of Watt to power the entire smart city. A set of weighted reward $R_{funders}$ is received by each *Funder$_i$*, per the amount of tokens of type *TokenType* within the vault, which represents a weighted linear combination of $S_E = \sum_{i=1}^{n} w_i S_i$ for some positive integer weights w_i. Each funder *Funder$_i$* submits his or her amount to the *Tokenized Energy Pool*. If the amount to be submitted *Stake$_{amount}$* is satisfied, the token generation begins based on the amount of money deposited on the *TradingVault* by the *Funders*. Once the funding is complete, a certificate is issued to participants within the market. The *Funding Contract* generates a set of unique IDs for each of the *Funders*. With this, they are able to access their funding rewards on the blockchain. As the tokens are minted and destroyed based on the various transactions in the Energy Market, *Funders* received a specific amount of interest i for which they can be able to either further fund the market or redeem the reward.

We then issue out a set of tokens known as *FundTokens*. This serves as a Reputation Tokens, which provides a unique sort of access as *Funders* of the blockchain energy system. This is to incentivize participants on the network to ensure the sustenance of the production of RE on the network with access to benefits b in a transaction. A percentage of token rewards *rewardtoken* is given to the *Funders* based on the trading that goes on at the Energy Market. All the rewards for the funders are deposited in the Funder Wallet *Fund$_{wallet}$*.

We also set a set of functions *balance()* to be able to track the number of energy tokens for each energy type within the *TradingVault* at each time t for each token type. Thus, at each time on the blockchain, we can know the number of reserves R of the various tokens within our smart city that balance with the amount of energy being produced in total to the entire blockchain-based energy smart city.

Upon token creation, the blockchain calculates the number of tokens needed to balance the entire blockchain-based smart city for each token type using the Automated Maker Pricing Mechanism. For the tokens *TokenTypes* created and approved within the *TradingVault*, we set a set of operations that the various token types can perform. Other information such as the name, the unique token symbol, the total supply, the balance, approval mechanism set, allowance module and transfer methods are logged into the blockchain. We set a set of functionalities to query this set of information on the blockchain at each point in each time. Thus, we are able to understand what is happening in the blockchain-based smart city. The token requesters (that is, the *EnergyBuyers*), *EnergySuppliers*, *DSOs* and *Energy Funders* entities within each module have particular tokens they can access and control for whatever tokenized transactions taking place on the blockchain. The entity can query these tokens from the token list generated on the blockchain.

Since Energy Market, several Energy Markets are thus created this way and funded on the blockchain with full access and monitoring; we observe an Energy Market A and B created that are connected together, with several corresponding *TokenizedEnergyPools* connected to each other and seen on the blockchain, in case of any change in reward R_{funder_i}, mispricing occurs between the two markets is created leaving an energy-backed arbitrage to be created for the two markets. If the reward R_{funder_i} is higher than the weighted sum of the prices of R_{funder_i}, then the arbitrageur can purchase R_{funder} by a profit price of $\left(R_{funders}\right) - P_{i=1}^{n} w_i$. Since the rewards incentivize Tokenized Energy Pools funders, we calculate the returns obtained for all the various energy types from the reserves kept within the *Tokenized Energy Pool Contracts* for the entire Energy Market, which generates the tokens for the various energy types. We take a scenario of two energy types α and ι.

Theorem 1
For each energy token supplied, increase and decrease in the token amount remains constant if relative return δ^t depends only on relative ratios of R_α and R_β.

Supposing we have a reserve of $R^t \alpha$ and we let $m_p^t \varepsilon R_+$ be the price of energy type α, respectively, at each time $t = 1, \ldots, T$. Using the Constant Product Function, we obtain $R_\alpha^t R_\beta^t = k$ for all t in the case of no arbitrage between Energy Market A and B, that is, $m_p^t = R_\beta^t / R_\alpha^t$ we combine both statements as $R_\beta^t = \sqrt{km_p^t}$. Expressing a relative return using the *PricingContract* on the blockchain between t_1 and t_2, we obtain

$$\delta^t = \frac{m_p^t R_\alpha^t}{m^t - 1_p R^t - 1_\alpha + R^t - 1_\beta} = \frac{R_\beta^t}{R^t - 1_\beta} = \sqrt{\frac{m_p^t}{m^t - 1_p}}$$

We extend this to the total relative gain for the *TokenizedEnergyPool* as

$$\delta = \prod_{t=2}^{T} \delta = \sqrt{\frac{m_P^T}{m_P^1}}$$

We thus achieve a total value of

$$P_v = \left(m_p^1 R_\alpha^1 + R_\alpha^1\right)\delta = 2\sqrt{km_P^T}$$

We thus prove that the results hold even when tokens are removed or added to the reserve from the transactions across the energy network.

10.3.5 Bidding on Energy Token Type

Various *EntityIDs* bid on the energy tokens within the *TradingVault* in order to both gain a certain amount of energy $energy_{amount_a}$ and exchange a certain amount of energy $energy_{amount_b}$. Ensuring that it is easy to identify various token types, we make sure the token types are easily discoverable via their names, symbols and token ids. Also, we provide a list of all tokens set to *Available* on the smart contract, and we display them to all entities on the blockchain smart city [10]. The bidding process works differently for the various energy token types based on the kind of tokens being bid upon. We set our tokens into two categories: *Fungible* and *NonFungible*. If the energy token types are set as *NonFungible*, only one bid is allowed for their exchange in the energy smart city. If the token type is set as *Fungible*, then that particular energy type with a particular amount of reserves *TokenType* can be bid upon by various entities and parties. *NonFungible* has a quantified amount of energy *Qenergy* that they match with. Once a person purchases a token of type *NonFungible*, he or she obtains a particular energy type of quantity *Qenergy*. The blockchain keeps track of the bidding session for the tokens being purchased by a particular entity *EntityID*. The blockchain records the amount of tokens requested as well as the particular *EntityID*, price P submitted, the current price of tokens *CurrentPrice*, equivalent amount of token types current token in reserve $R_{tokenamount}$, Eq_{token}, and Amount of energy Eq_{Watt} to be supplied to *EntityID*. After each bid, the balance function recalculates the value of the remaining tokens and broadcast it to the blockchain. Thus, the current price of the tokens at each time on the blockchain is easily determined. We ensure that the owners of the tokens such as *DSOs* cannot be able to spend the tokens that have been bid upon.

10.3.6 Blockchain-Based Energy Market Token Pairing

We set up a pairing mechanism for the various token types with all tokens in the vault through the *Pairing Contract*. Several transactions are permitted to take place. For instance, *Tokenized Energy Pool* provider can pay fees as well request for the submission of a specified amount of token types that are paired with their address *Tokenized Energy Pool* provider address. Participants on the chain can also make Token-Token Request *TTR* of several energy types. In some cases, participants are rewarded with Reputation Tokens along with energy tokens. These fees are transferred to the contract address. Upon request, the Pairing Contract calculates the number of energy tokens that will be enough to be submitted per the current-voltage level in the market. It also takes into account the amount of energy to be supplied or demanded based on the request of participants in the smart city. It sets the amount to be paid by the particular participant requesting f_{in}. It then sets the corresponding amount of energy needed to be submitted per the current policy and regulations set by the *Pairing Contract* as *AmountIn*. We also calculate the amount of tokens needed to meet the demands as *AmountOut*. We set a corresponding fee to be paid by the particular entity, making the transactional request as f_{out}, which is produced by the P based on the current estimations. The *Pairing Contract* contains rules of how one type of token will be traded for another to control the energy use in the entire smart city. We ensure that the vault and the market are balanced by using the CPM Function. If the request sent submitted requires the reduction of a particular energy type, the tokens of the specified energy type *EnergyType$_i$* are destroyed. This reduces the amount of tokens for type *EnergyType$_i$*, which meets the current amount of energy within the Energy Storage Module. We adjust the *TradingVault* to meet the current changes in a process known as rebalancing. Different energy types can also be paired across other energy markets as long as their exchange function is equivalent.

10.3.7 Rebalancing of Tokens Generated

Several token types are generated, each serving different functionalities on the blockchain. The aim of all these tokens generated is to incentivize behaviors that ensure the effective management of energy and effective distribution of energy across all *DERs* and entities found within the various modules on the network. It is important for rebalancing to occur to ensure that all energy types within the *TradingVault* are set to a fixed allocation of *token$_{alloc}$*. Balancing of tokens on the Energy Market is ensured as far as there are payments of f and various fluctuations of prices on the network. However, in the event where $f = 0$, all the benefits of rebalancing on the Energy Market disappears. As such it is important for *Tokenized Energy Pool* to remain constant in the event where $f = 0$, we ensure [4, 10] that the reserves for the various energy type is kept at $R_\alpha R_\beta = C$ where C is the geometric mean of all balances of all energy tokens $R(R_i, \ldots, R_n)$ the quantity that must increase in order for the rebalancing to occur. We observe a growth in the geometric means as quantities grow. As such, we implement an f system to ensure balanced reserves of our energy token and balance the Energy Market and effectiveness of the amount of energy generated and distributed across the blockchain network.

10.4 Energy Trading

Since the tokens provided work along with multiple smart contracts to control the entire smart city, it is easy to manage the multiple entities and incentivize all participants within the network to behave according to the rules

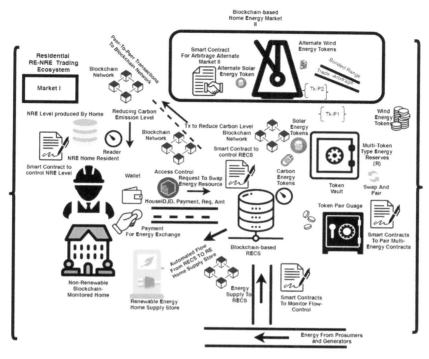

FIGURE 10.5
Blockchain-based trading architecture for complex NRE/RE smart city over automated Energy Market maker.

set. The blockchain more effectively manages the complexity of the various transactions occurring simultaneously due to its decentralization as shown in Figure 10.5.

10.4.1 NRE/RE Energy Market Trading

We set up several Tokenized Energy Pools with multiple markets. We set up an *Auxillary Market* from another smart city. We set the two Energy Markets as Market *A* and Market *B*. Upon request received by the *Residential Entities* based on demand *D* from the blockchain, generate the specified number of tokens based on the token generation process. The tokens generated are kept in a *TradingVault*. Each token type is set into an individual address allocation from which they are transmitted upon request. Several token requests are received. With several token types kept in the Energy Market Vault, we better develop an exchange mechanism to enhance energy balancing across the entire smart city. Since the exchange of a particular energy type, *A* affects the amount of energy type *B* exchanged, there is a need to balance the entire smart city through pricing and balance.

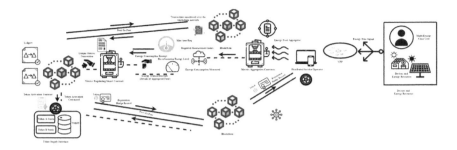

FIGURE 10.6
A tokenized energy generation mechanism and control for multiple NRE/RE entities.

We first set up a limit of l for each token representing a particular amount of energy type $EnergyType_i$ on the blockchain. Each token of a particular energy type has a specified amount n that can be minted on the blockchain. Thus, the amount of energy type $EnergyType_i$ that can be sent into the grid is capped based on the present limit of tokens amount $token_{amount}$ that can be minted on the blockchain. Thus, the amount of energy that can be sent to the *EnergySupply* market is thus regulated in this way. This also helps maintain a balance in the voltage distribution across the network.

A trading system is created as shown in Figure 10.6 within the Energy Market to exchange energy types. Entities on the network requesting and distributing energy tokens begin to initiate a request to the blockchain-based Automated Energy Market Maker *AEMM*, which underlies the Energy Market. Entities may want to exchange various tokens for other tokens in very complicated ways. We set the output request *Energy Output*, which is made on the market as $\Lambda \varepsilon R_+^n$, which is expressed as a vector with nonnegative entries. The *ith* entry specifies how much energy tokens the *EnergyBuyer* will like to receive from the Energy Automated Market Maker *EAMM* for an equivalent amount of *NRE/RE* energy supplied. We also define the input $\Delta \varepsilon R_+^n$, which is a vector whose *ith* entry shows how much energy has been supplied to the *EnergyMarket* and the number of new tokens generated by the *AEMM*. From this, we set the exchange *exch* as $\Delta \Lambda \varepsilon R_+^n * R_+^n$.

10.4.2 InterEnergy Type Token Request and Transaction

A trade function φ is then specified exactly when a trade is considered valid. A trade is only accepted by the *TradingContracts* only when Δ, Λ satisfies $\varphi(R, \Delta, \Lambda) = \varphi(R, 0, 0)$. This means that the trade is only accepted as long as φ is kept constant. From the *TradeContract*, the *EnergySupplier* then supplies the energy of amount $s\Delta p$ the energy tokens supplied for an equivalent which the *EnergyBuyer* takes. The reserves of the energy type are then updated as $R \leftarrow R + \Delta - \Lambda$. The energy exchange can be either of type *NRE* or *RE*. If it is of type *NRE*, the equivalent amount of energy is supplied to the NRE as requested. If it is of type RE, the energy is supplied from the *RECS* to the *RenewableEnergyHomeSupplyStore*. In our work, we set a restriction for entities to make a request for *NRE* energy token types and reduce demand for *NRE* tokens generated; this is to incentivize entities on the blockchain-based energy smart city to change to *RenewableEnergyType* and increase the amount of RE generated in the smart city. Apart from this, *NRE* entities can request any other token types, such as Reputation Tokens and other types of tokens. *NRE* token types can be generated and only exchanged for *RE* tokens. The NREs generated in the smart city are destroyed upon exchange. The relevant payment is made to the *EnergySupplier*. This provides a means to reduce the amount of *NREs* that is produced across our blockchain-based energy smart city. Due to the competitive demand involved in trading the energy tokens between *EntityBuyer* and *EntitySupply* at a fee of $1 - \gamma$, each entity attempt to obtain a greater payoff. Since we have n number of token types within the *EAMM*, the energy exchange function is $\varphi(R, \Delta, \Lambda) = (R_i + \gamma \Lambda_i - \Lambda_i)(R_2 + \gamma \Delta_2 - \Lambda_2)$.

Although as frequently in the Energy Market, there is a frequent cause of non-zero fees. However, it must be noted that no rational entity within the smart city will ever opt to make both ($\Delta_i \neq 0$ and $\Lambda_i \neq 0$) in the case of non-zero fees. This is to say that no particular entity will exchange a particular energy token type for a smaller amount of the same token type Δ_2 and Λ_2. The same is the same for the energy token pair. We express this behavior in the following equation:

$$\varphi(R, \Delta, \Lambda) = \begin{cases} (R_1 + \gamma \Delta_1)(R_2 - \Lambda_2) & \Delta_2 = \Lambda_1 = 0 \\ (R_1 - \Lambda_1)(R_2 + \gamma \Delta_2) & \Delta_1 = \Lambda_2 = 0 \\ \varphi(R, 0, 0) + \varepsilon & \text{otherwise} \end{cases} \quad (10.4)$$

10.4.3 Feasibility of Energy Exchange Across Energy Markets

Again, to ensure that all the entities stay within a particular Energy Market, we must ensure that it is cheaper to trade in that Energy Market. However, entities will still want to exchange energy token types across other Energy Markets. It is thus important to set a relation between these two Energy Markets and the specific energy type being exchanged to balance the energy generated within the smart city since changes in energy cause specific changes in voltage within the smart city. A exchange (Δ, Λ) is feasible across two or Energy Markets if their trading set $(\Delta, \Lambda)\varepsilon T(R)$ is equivalent. That means if two energy sets are not equal $T(R) \neq T'(R)$, then there exist a trade set that is not feasible in the other. One way to ensure that is to compare the exchange function, φ and $-\varphi$ from the two *Energy Market Maker*. We thus ensure that the two Energy Markets are equivalent by setting the functions of both to zero.

$$T(R) = (\Delta, \Lambda) \mid \varphi(R, \Delta', \Lambda') = 0 \quad \text{for some} \quad \Delta' \leq \Delta, \Lambda' \geq \Lambda$$

Because the weighted geometric mean function is a concave function, we see a set of convex trade sets on the blockchain. This indicates that the trading system is in place.

$$T(R) = (\Delta, \Lambda) \Big| \varepsilon R_+^n * R_+^n \Big| \prod_{i=1}^{n} (R_i + \Delta_i - \Lambda_i)_i^w \geq \prod_{i=1}^{w} (R_i)^{w_i}$$

is therefore a convex set, for any valid weights $w \geq 0$ with $1^{T_w} = 1$ and reserves R.

10.4.4 Energy Storage and Supply for Trading

For trading among multiple entities in the market, we check the amount of energy stored within the Energy Storage and ensure that it balances with the energy stored up within the market. This energy is supplied by the prosumers within the blockchain smart city who provide the various energy types along with generators and suppliers. This forms the core of the *EnergySupplyUnit*. They serve as a marketplace for the selling and purchase of surplus RE on a local level. Through their microgeneration devices, they generate a fraction of the energy they consume. Due to the nature of the blockchain, which ensures a P2P means of transactions, amount of prosumers (producers and consumers of energy resources) risk is shared among multiple parties. With the negotiation contracts set between the *Energy Market* and the prosumers, a far agreement is set in the amount of energy that prosumers are willing to provide into the Energy Storage System. We set up a tokenized system for the Energy Supply Unit to incentivize participants to supply the correct amount of energy on demand. This control we provide a control system to regulate the requested amount of energy Amt_D and the quantity supplied Q_s. This is controlled by the *EnergySupplyContract*. This interfaces between the Energy Storage System and the *Energy Supply Unit*. If the parties are making the request, for instance, the prosumers do not comply with the requirement *req* set within the contract, the requesting device is penalized with a set fee f_p. Before initiating the request for energy by the energy suppliers, to the storage, the storage device pays an amount of fees $f_{storage}$. The amount is based on the amount requested from the market to balance the entire smart city and the rate of transactions across the Energy Market at time $\$t\$$. This affects the amount of energy being requested by the Energy Storage Unit Amt_D being made to the Energy Supply Unit. The Energy Storage is powered by *EV Owners*, *Power Companies* and *Storage Providers*. They provide a means to be able to supply energy units, storage credits and reward tokens to entities supplying energy within the smart city.

The Energy Storage is supported by *EV Batteries* to add to the storage capacity. This regulates the optimal amount of energy to be supplied to the blockchain-based Energy Storage and the exchange and also manages negotiations among generators and suppliers. As such, the amount of energy within the Energy Storage is system is of adequate amount. The amount of energy supplied by a particular prosumer is calculated as a percentage of the amount of supply toward the entire blockchain-based smart city. The price of the energy type being supplied is known at the time the energy supply is sent across the network at time t_{supply}. Due to the continuous transactions taking place at *SupplyTransaction* and the continuous demand D offered by both buyers *EntityBuyers* of a particular energy type as well as *NRE* users trying to reduce their energy amount, there occurs a fluctuation of prices p across the network creating a complex set of demands. Also, due to blockchain's ability to provide complex and transparent instructions for multiple nodes on the entire network, uncertainty is reduced for the loads of prosumers and generators with their energy resources since it is easier to make a set of decisions $[D_i, \ldots, D_n]$ at a certain time on the blockchain. The *NegotiationContract* provides a fair means of interactions between parties supplying energy to the blockchain-based Energy Storage System.

10.4.5 Negotiation Analysis between RE/NRE Buyer and Supplier

We make use of incentive theory to understand the number of tasks needed for a single energy supplier such as a prosumer to supply the various entities on the network for our *negotiation contracts*. We assume that each supplying entity *EntitySupply* is facing a multi-tasking problem based on several requests Req_{demand} coming from the *Energy Market*, in a scenario of k requests being sent to the *EntitySupply$_i$*, has to allocate activity $y \varepsilon Y = \{y^1, y^2, \ldots, y^k\} | \{y^1 + y^2, \ldots, y^k \leq T\}$ among a number of different tasks in response to the cost and benefits of the different tasks. We set T as *EntitySupply*, the total time available to allocate between tasks. We observe a cost function $C(y, \beta) = \Sigma k_{j=1} c(y^1, \beta^i)$ where $c(y^i, \beta^i)$ refers to cost of allocating all relevant efforts to the demand i. If y^i is zero, then we resolve the cost as zero else, we set the cost as $\beta_i y_i^2 + f$ on the blockchain. The cost parameter β^i is a random variable that can take one of m discrete values. The benefit received by the energy supplier is assumed to take the form $B(y, \alpha) = \alpha^T y$. From the *Negotiation Contract*, the smart contract sets the term of agreement between our *EntitySupply$_i$*, and the *EntityBuyers*, the buyer's preferences are represented by the utility function $u(q, T, \theta) = \int_q^0 P(x, \theta) - T$ where q is the number of units purchased, T refers to the total amount to be paid to *EntitySupply$_i$*, and $P(x, \theta)$ is the inverse demand curve of a buyer with preference characteristic θ. We thus consider the following functional form for *EntityBuyers* preferences as $u(q, T, \theta) = \theta v(q) - T$. However, in the negotiation process, characteristics of θ are known to all parties transacting on the blockchain. The *EntitySupply$_i$* also knows the distribution of $\theta, F(\theta)$. Assuming the cost of sending a unit of energy by the *EntitySupply$_i$* is given by $c > 0$ his payoff from selling q units T against a sum of the fee paid T is given by $\pi = T - cq$.

We try to understand the profit-maximizing pair *(T,q)* that *EntitySupply* will be able to induce *EntityBuyers* to choose. We realize that maximizing information can only be achieved based on the awareness of necessary information, which is made possible by the transparency of the blockchain [15].

One disadvantage for *EntitySupply*, in the setup used in this work, is that *EntityBuyers* has several options to request for several energy types due to the various numbers of *Tokenized Energy Pools* he has access to. Thus, it is important for *EntitySupply* to move away from linear pricing. However, we will be able to do so if the buyer cannot make an arbitrage on other *TokenizedEnergyPools*. However, in the event of *EntityBuyers* being able to able to trade in other *Tokenized Energy Pools*, and arbitrage being costless, we retain linear pricing as the only possibility since that *EntityBuyers* can be able to buy a minimum average price and then resell them at the alternative *Energy Type Market B* or *ETLP$_B$*.

We show that *EntitySupply* can do better by offering more general nonlinear prices. Since the *EntitySupply* knows the type for *EntityBuyers* based on the transparency of information on the blockchain, with the use of our *PricingContract*, we can be able to optimize the negotiation under an agreed schedule using static bilateral contracting between the two *EntityBuyers* and *EntitySupply* where L and H refer to the two buyers involved. Each buyer has, in this case, certain preference characteristics $\theta\varepsilon\theta_L\theta_H$ that all parties know on the blockchain with a probability of $\beta\varepsilon[0,1]$ and $1-\beta$, respectively. We set β as the proportion of *EntityBuyers* of type *NRE/RE*.

Now at the set schedule *T*, *EntityBuyers* requesting for an amount of energy $\left[q, T(q)\right]$, we pick the outcome that maximizes his payoff. *EntitySupply* thus solves for

$$\max_{T(q)}\beta\left[T\left(q_L\right)-cq_L\right]+(1-\beta)\left[T\left(q_H\right)-cq_H\right]$$

subject to

$$q_i = \arg\max_q \theta v\left(q\right)-T\left(q\right) \text{ for } i=L,H$$

and

$$\theta_i v\left(q_i\right)-T\left(q_i\right)\geq 0 \text{ for } i=L,H$$

We achieve π^* where *EntitySupply* is able to meet the optimizing quantity q_i for both buyers under the agreed schedule *T* given the incentive constraints i_c in the Energy Market.

10.4.6 Energy Token-Price Fluctuation Analysis

Due to the constant exchange of various energy types, fluctuations in prices result in drastic changes in the amount of tokens within the energy type.

The blockchain provides a means of control through its transparency and self-verification mechanism within it. We set a specific price p for which the price of each token must fluctuate around. We set a bounded range of $[1-\varepsilon, 1+\varepsilon]$ for some $\varepsilon > 0$ using our *PricingContract*. The fluctuation of these token types is dictated by their natural sources of demand and can vary greatly. The blockchain provides a means to match the physical quantity of these energy tokens with their token quantity and pricing within the Energy Market. Again, due to the varying *TokenizedEnergyPool*$_1$, ..., *TokenizedEnergyPool*$_n$ for various energy type, fluctuations do occur in with *TokenizedEnergyPool*.

We analyze the changes in pricing that occur as the token fluctuates within the *TokenizedEnergyPool*. Before the *EnergyBuyer* from the residential entity module makes a request to the Energy Market through the *EAMM*, we set the time to $t = 0$. We compare the current prices of the Energy Type Tokens to the prices on other Energy Markets, which are broadcasted all across our blockchain-based smart city. The trader undergoes this by trading an amount of an energy type ΔX_t for $\Delta^* Y_t$ at a particular time t. Based on the difference in prices based on the difference in demands across the various Energy Markets, the *EnergyBuyer* makes a transactional request that maximizes his profit. However, this is controlled based on the restrictions on the fees (f) to $1-\gamma = {}^*\gamma$, not $(1-\gamma)$, which forces a change in the price of a particular token type by a factor of e^δ or a factor $e - \delta$. We utilize a price curve to know the current price and demand changes as time evolves in the Energy Market. We observe the following conditions by which the prices $\$S\$$ of energy tokens are affected.

If we are $t = 0$, we have the prices of our tokens set to $S_0 = X_0 = 1$. At time t, the price is set at $S_t = e^{(k_y+1)\delta} S_t^0$. At this point, the price is exchanged at a quantity $\Delta X_t = X_t \left(\dfrac{\delta}{e^\gamma + 1} - 1 \right)$ for $\Delta Y_t = Y_t \left(1 - e \dfrac{-y\delta}{\gamma + 1} \right)$ of energy type tokens. The implicit price of the tokens after each transaction is set to $S_t^* + \delta_t = \gamma^{-1}$ on the price curve and broadcasted to all entities on the network.

At another particular time, the price is set as $S_t = e^{-\delta} \gamma S_t^*$. The asset is exchanged at a quantity $\Delta Y_t = Y_t \left(e \dfrac{y\delta}{\gamma + 1} - 1 \right)$ of a particular energy type for $\Delta X_t = X_t \left(1 - e^{\frac{y\delta}{\gamma+1}} \right)$. The implicit price broadcasted across both Energy Markets after the exchange is $S_t^* + \Delta t = \gamma S_t$. However, there we maintain a standard price on the *AEMM* in the Energy Markets and thus profitability is possible in exchanging across the markets.

At the next period t, we set time as $e - k_y \delta S_t^* \leq e^{k_y^\delta} S_t^*$. At this point, there is no interaction between the *EAMMs* since no amount of exchange of token type is profitable.

Using Markov Chains, we can understand each point of the blockchain where the exchange of energy-type-token occurs. We observe that at a particular time on the Markov Chain, with $M_t = \log \dfrac{S_t}{S_t^*}$ and a state-space on our blockchain network, a trade occurs each time the Markov Chain stays at one end of the two end states.

Using the blockchain and our *PricingContract*, it is not hard for us to verify if the exchange on the markets has a stationary distribution.

$$\left[\frac{1}{2k_\gamma+1}, \dots, \frac{1}{2k_\gamma}+1\right].$$

For us to be able to obtain the number exchanges N_t occurring on the Energy Market at time t, in the absence of a fee paid, we calculate the payoff in the $K_\gamma - N_t/2\left(Y_t S_t^* + X_t\right) = \left(X_t, Y_t\right)^{\frac{1}{2}}\left(Y_t S_t^* + X_t\right)$ *EAMM* as where refers to the payoff within our *EAMM*. Finally, to fully understand the asymptotic geometric return of our energy token pool, we set our time to a limit T. We express this as

$$\lim_{T\to\infty}\frac{1}{T}E\left[\log W_T\right] = \lim_{T\to\infty}\frac{1}{T}\left(E\left[\log\left(Y_T S_T^* + X_T\right)\right] + O(1)\right)$$

$$= \lim_{T\to\infty}\frac{1}{T}E\left[\log K_\gamma^{\frac{N_T}{2}} + \frac{1}{2}\log S_T^*\right]$$

$$= \lim_{T\to\infty}\frac{\log K_\gamma E\left[N_T\right]}{2T} + \lim_{T\to\infty}\frac{1}{2T}E\left[\log S_T^*\right] \tag{10.5}$$

$$\lim_{T\to\infty}\frac{\log K_\gamma E\left[N_T\right]}{2T} + \frac{n\delta(2p-1)}{2}$$

Theorem 2
Given a set of nodes n on the blockchain, it is possible to obtain the average number of exchange transactions using our set of contract C within Energy Automated Market Maker M.

Since we have the stationary distribution as mentioned previously, and limit of $\lim_{T\to\infty}$, it is possible to obtain the average number of exchange transaction using our *PricingContract* within our *EAMMs*. We thus compute the average number of exchange of energy type tokens being transacted at an interval of time T as $nT\left((1-p)\pi\left(-k_\gamma\delta\right)+\rho\pi\left(k_\gamma\delta\right)\right)$ using our *PricingContract*.

$$\lim_{T\to\infty}\frac{E[N_T]}{nT} = \begin{cases} \dfrac{1}{2k_\alpha+1} & \text{if } p=\dfrac{1}{2} \\[3ex] \left(1-\dfrac{1-p}{p}\right)\dfrac{p+(1+p)\left(\dfrac{1-p}{p}\right)^{2k_\alpha}}{1-\left(\dfrac{1-p}{p}\right)^{2k_\alpha+1}} & \text{if } p>\dfrac{1}{2} \end{cases}$$

Token Types	Units Accessing	Token Type	Users Access
Reputation tokens	Energy market, non-renewable energy residence module, renewable energy residence module	Fungible Tokens, Non-Fungible Tokens	Energy traders, DSOs
Energy supply tokens	Energy storage systems, Energy Supply Module	Fungible Tokens	EV batteries, solar charging batteries, renewable energy residential home energy units
Energy type tokens	Energy market	Fungible Tokens, Non-fungible Tokens	Energy traders, residential units, industrial units
Power generation tokens	Energy Market, Energy Supply Module, Energy Storage Module	Fungible Tokens, Non-Fungible Tokens	Wind distributed energy devices, solar energy devices, carbon energy devices, energy traders, residential units, industrial units
Energy exchange tokens	Energy Market, Non-Renewable Energy Residence Module, Renewable Energy Residence Module	Fungible Tokens	Industries, non-renewable energy residence

10.5 Conclusion

In conclusion, we designed a tokenized blockchain-based energy smart city that transacts over Automated Energy Market Makers. We provided various entities producing various energy types to co-exist in a single smart city. We designed a mechanism for incentivizing participants within the NRE smart city to decarbonize and exchange energy between various RE entities within the same smart city. In structuring the smart city into residential, Energy Storage, market and supply, we designed a novel scheme where tokenized systems can be used to regulate energy systems in a decentralized autonomous manner over the blockchain ensuring energy efficiency.

References

[1] Wang, Q., & Su, M. (2020). Integrating blockchain technology into the energy sector—from theory of blockchain to research and application of energy blockchain. Computer Science Review, 37, 100275.
[2] Mezquita, Y., Gazafroudi, A. S., Corchado, J. M., Shafie-Khah, M., Laaksonen, H., & Kamišalić, A. (2019, October). Multi-agent architecture for peer-to-peer

electricity trading based on blockchain technology. In 2019 XXVII International Conference on Information, Communication and Automation Technologies (ICAT) (pp. 1–6). IEEE.

[3] Clark, J. (2020). The replicating portfolio of a constant product market. Available at SSRN 3550601.

[4] Evans, A. (2020). Liquidity provider returns in geometric mean markets. arXiv preprint arXiv:2006.08806.

[5] Cejka, S., Einfalt, A., Poplavskaya, K., Stefan, M., & Zeilinger, F. (2020, September). Planning and operating future energy communities. In CIRED 2020 Berlin Workshop (CIRED 2020) (Vol. 2020, pp. 693–695). IET.

[6] Camacho, E. F., Samad, T., Garcia-Sanz, M., & Hiskens, I. (2011). Control for renewable energy and smart grids. The Impact of Control Technology, Control Systems Society, 4(8), 69–88.

[7] Dickerson, F. B., Tenhula, W. N., & Green-Paden, L. D. (2005). The token economy for schizophrenia: review of the literature and recommendations for future research. Schizophrenia Research, 75(2–3), 405–416.

[8] Karandikar, N., Chakravorty, A., & Rong, C. (2021). Blockchain based transaction system with fungible and non-fungible tokens for a community-based energy infrastructure. Sensors, 21(11), 3822.

[9] Voshmgir, S. (2020). Token Economy: How the Web3 reinvents the Internet (Vol. 2). Token Kitchen, Berlin, Germany.

[10] Angeris, G., & Chitra, T. (2020, October). Improved price oracles: constant function market makers. In Proceedings of the 2nd ACM Conference on Advances in Financial Technologies (pp. 80–91). Association for Computing Machinery, New York, NY.

[11] Mengelkamp, E., Notheisen, B., Beer, C., Dauer, D., & Weinhardt, C. (2018). A blockchain-based smart grid: towards sustainable local energy markets. Computer Science-Research and Development, 33(1), 207–214.

[12] Shahsavari, A., & Akbari, M. (2018). Potential of solar energy in developing countries for reducing energy-related emissions. Renewable and Sustainable Energy Reviews, 90, 275–291.

[13] Zenginis, I., Vardakas, J. S., Echave, C., Morató, M., Abadal, J., & Verikoukis, C. V. (2017). Cooperation in microgrids through power exchange: an optimal sizing and operation approach. Applied Energy, 203, 972–981.

[14] Thomas, L., Zhou, Y., Long, C., Wu, J., & Jenkins, N. (2019). A general form of smart contract for decentralized energy systems management. Nature Energy, 4(2), 140–149.

[15] Tassy, M., & White, D. (2020). Growth rate of a liquidity provider's wealth in xy= c automated market makers. Available: https://math.dartmouth.edu/~mtassy/articles/AMM_returns.pdf

11

Smart City Governance

11.1 Introduction

With the advancement of new technology, the concept of a digital city is gaining traction as a means of creating more efficient and sustainable cities. Cities are becoming smart not only in areas such as automation and daily functions but also in ways that allow planners to comprehend, monitor, evaluate, and plan the city in real time to improve its performance. The importance of environmental and social capital distinguishes digital cities from a technology-centric concept that isn't blended, resulting in a more multidimensional picture of cities [1]. Cities planners must govern their development in order to become smart by bolstering economic competitiveness, increasing social unity, assuring environmental sustainability, and ensuring a better quality of life for their residents.

According to a recent survey, 1.5% of the world's population lives in cities. This puts a lot of strain on the resources of cities. Digital cities have emerged as the ideal answer for addressing the difficulties that face metropolitan cities, which are intricately linked to each city's social, political, cultural, and territorial networks. The definition of digital cities supports the notion that it is a multi-stakeholder, municipal issue. As a result, governance issues become a key worry, making it difficult for many stakeholders to collaborate in the creation of digital cities.

11.2 Urban City Challenges

In this section, we will look at some of the governance issues that face urban areas.

11.2.1 Governance Challenges

The issues of governance in urban areas are inextricably linked. Institutional instability, limited capabilities, and the distance between them and the

DOI: 10.1201/9781003289418-11

population are all concerns that call for a change in governance models [2]. The existing link between the government and the governed must be strengthened through tools such as participation, as well as the recognition of residents' needs and the coverage that social services should give. The spatial distribution of government and city growth drivers has emerged as a critical issue that cities must address.

1. *Low capacities of the urban institutions*: Governance systems in urban cities present weaknesses in formal institutions. Therefore, there is the need for adequate systems of usage of lands, registering of activities, data collection, and proper management of houses and markets. Improvements in city administration services are required, as well as the flexibility to respond to public requests with the required speed and agility [3].

2. *Lack of stability in the system of governance*: Governments all around the world are attempting to implement various democratic governance models in order to overcome the opposition of existing political and economic interests, as well as undemocratic schemes. Cities may be impacted directly or indirectly by the new democratic spaces that have been created as a result of their importance to this agenda.

3. *Existence of gap between the govern body and the government*: There is the existence of gap between inhabitants and decision makers regarding political, cultural, and economic aspects, and this must be looked at. The inhabitants want to have in order to access and manage resources but well-established systems make these differences last in time.

4. *Too much centralization and lack of coordination among institutions*: The division of power between the national government and the local government is unequal in many cities. Municipalities, for example, lack the ability, authority, and jurisdiction to take autonomous action on smart city projects. Lack of institutional coordination is a problem at multiple scales, including between different levels of government (national, regional, and municipal authorities), as well as between cities.

11.2.2 Economic Challenges

In terms of the economic dimension, the key concerns in metropolitan regions include unemployment, resource availability, and other competitiveness issues. Balanced urban productive tissues, on the other hand, are critical components for facing economic challenges and promoting city competitiveness from a geographical standpoint.

1. *Economy weakness and low competitiveness*: It is critical to foster a competitive and open business environment, as well as fair access to funding and job or business prospects. For improving domestic

productivity and revenue generation, the economy of urban cities must be self-driven by leveraging existing strengths, management innovation, and new technologies.

2. *Lack of diversification on urban economy*: Urban economies are typically concentrated in one or a few sectors, reducing their resilience. Economic diversification, which is the process of changing an economy away from a single revenue source toward many sources from a widening range of industries and markets, is one of the main components for enhancing productivity and employment generation in metropolitan regions [4].

3. *Excessive weight of informal economy*: Because of its widespread presence, the informal sector plays a significant role in the economy of metropolitan areas, as well as the opportunities it provides for the less educated and disadvantaged. Even if it cannot be completely managed, it must be looked at when addressing issues relating to economic challenges for urban areas.

4. *Shortage in access to technology*: The majority of city residents have access to basic utilities such as piped water, sewerage, and electricity. However, in metropolitan areas, the percentage of people with access to communication technology is lower than predicted. In addition, digital education and skill development for residents and professionals are required.

11.2.3 Mobility Challenges

It is vital to have a systematic approach to mobility concerns in relation to the overall difficulty that metropolitan regions face. To alleviate the challenges of public transportation, infrastructure deficits, and population concerns in the city, a more holistic approach through urban planning and mobility systems is required. Accelerated urbanization processes pose a serious threat to these goals, and they must be addressed.

1. *Lack of accessible and affordable public transport*: It is critical to fund inclusive public transportation networks that allow for connectivity and convenience throughout the city. New modes of transportation and regulations must be implemented in order to promote change toward more sustainable and equitable mobility solutions.

2. *High infrastructure deficit*: In metropolitan regions, there are general infrastructures; nevertheless, networks are in poor condition and are unable to provide the requisite capacity to meet the ever-increasing demand of the population. As a result, promoting the renewal and enhancement of networks for mobility, ICT networks, water supply and treatment, effective healthcare, and so on is critical.

11.2.4 Pollution

Water pollution and air pollution are serious issues in urban areas that have a negative impact on human health and the environment.

1. *Challenges in the environment*: Environmental concerns in metropolitan settings are exacerbated by inefficient resource management and pollution. The scarcity of resources and the effects of climate change are the two most pressing issues confronting metropolitan regions. The fundamental causes of environmental difficulties in cities are the spatial patterns followed by city development, in which speed and absence of controls, climate change effects, and inequalities in the geographical distribution of cities all play a crucial role.

2. *The effect of climate change challenges*: The rise in global temperatures is projected to result in desertification and water-related issues. Climate change may result in a reduction in access to essential human necessities. Natural calamities such as droughts, floods, and harsh weather must be considered while planning urban designs.

3. *Lack of fairness in access to opportunities and resources*: There are gaps and discrepancies across demographic groupings in terms of age, gender, origin, disabled individuals, and so on. The disparity results in unequal economic and social chances. Access to urban-based employment and housing is hampered by a large demographic share. To ensure equitable chances for all members of the community, the gaps must be narrowed.

4. *Very rapid urbanization*: The number of developing cities is rapidly increasing. Cities are currently rapidly expanding their boundaries, in contrast to the natural progression of traditional cities. The prevalence of slumps is a crucial component in the expansion of urbanization. Rural-to-urban migration and refugee influxes have exacerbated urban population pressures throughout time.

11.2.5 People Challenges

Unfairness, education, and culture are all issues that people face. Cities' challenges can be alleviated by reducing disparities and improving the quality of access to innovation and education. All activities taken to address these difficulties must take into account identity issues as well as the unique characteristics of cities' social urban tissue.

1. *Low educational level and digital skills*: There is a shortage of access to an enabling environment that allows for widespread access to education and consistent skill development. In order to maximize development potential, educational levels must be raised.

2. *High obstacles to social mobility*: Community groupings are strictly defined and impenetrable, impacting youth and new generations' visions of the future. This isn't simply a question of reality; it's also about the possibility of future change, and it's linked to migration, so it'll affect the future vision of the next generation.

3. *Urban poverty and inequality*: In some countries, there is a significant lack of equality. People who live in cities have advantages that people who live in countryside do not have. As a result, many individuals are migrating from rural to urban areas in pursuit of greener pastures.

11.3 Blockchain

Blockchain technology was first created to solve the two main issues inherent to digital currency. The first issue is double spending (a result of how easy it is to reproduce an identical copy of any digital file). This implies that, in digital financial transactions, the "copy paste" takes a sinister turn when the copied information is a monetary exchange value [5]. This promotes fraud and transfers of amounts that are not in existence. The next issue has to do with the need for a central authority to validate payments. Central authorities (Central banks) are the only agents with the power to issue currency and guarantee its authenticity. The transfer of money is controlled by a financial authority that ensures that the transfer of assets is correctly recorded. This connotes that we put our trust in the authority that issued the money and the reputation of the intermediary service.

Currently, new platforms on the Internet have resulted in a vast increase in everyday transactions and the risk of dealing with an unknown party. What guarantees that the intermediary service is efficient and meets our need for direct and safe transactions? How can we ensure that the currency is used just once in a transaction and, as such, it is not copied illegally to be used in other transactions? Satoshi Nakamoto introduced blockchain technology in 2008 to answer the previous questions Blockchain is a distributed ledger technology that permits a network of databases to develop. In these systems, individuals of a given community can create, approve, store, and share data securely [6] and proficiently, at any time and without geographic impediments. These systems work without a central overseeing body (or with one if they select) and can show the complete exchange history (or choose to keep it covered up). In any case, altering the points of interest of the transaction history, or a few of its records, is essentially inconceivable, suggesting the data stored there is profoundly solid. Each node offers a decentralized duplicate of the information stored on it and has access to the information (directed or otherwise).

This implies that blockchain democratizes the capacity to approve exchanges, which was already limited to central, regularly national, monetary frameworks. This innovation in this manner empowers modern environments to exchange budgetary and nonfinancial resources. Blockchain made Bitcoin conceivable, the first-ever encrypted digital cash, or cryptocurrencym [7]. Utilizing cryptography, Bitcoin recognizes the members on a network and encrypts the messages sent between them. It then employs consensus to construct an affirmed record of their interactions. This record is shared among all the individuals of the community to ensure it cannot be adjusted. It in this manner archives the possession and exchange of advanced resources with inalterable exactness. As computerized esteem, money recaptures its potential for sharing data and its use as a social instrument. In this manner, blockchain opens up modern opportunities for collective coordination, permitting cash to recapture its unique work as an apparatus for social interaction and as a mechanism for trading esteem.

11.4 The Role of Blockchain in City Governance

Blockchain's elementary level groundbreaking viewpoint lies within how it decentralizes agreement and permits unidentified parties on any network to believe each other for interactions and exchanges. Both of these issues lie at the heart of any administration framework, including the governance of cities. Blockchain's genuinely transformative nature lies within the decentralized agreement of belief in peer-to-peer interactions, without any need for approval from a central authority [8]. The innovation's immediate impact is to strengthen the capacity of specialists and citizens in a given area to coordinate successfully. Blockchain's permanent and shared exchange record gives straightforwardness and unquestionable status, which suggests its potential, will permit us to extend our information about cities in the future and effectively drive forward equitable forms that empower social consideration and success.

Blockchain offers the plausibility of trust, agreement, and information to move forward the viability and productivity of city governance. Cities confront significant financial challenges and natural dangers. Their governance is regularly divided between several diverse specialists with restricted powers, avoiding them from advertising a facilitated response to challenges, making it more challenging to construct social cohesion within the city [9]. Blockchain technology can create a modern institutional structure for city governance, as long as it is utilized to organize interactions among the parties involved in open issues, to make choices, and to draw up rules, or the laws of the game, that permit us to realize the best for society. Blockchain can offer assistance to execute a people-centered plan, an administration based on transparency, modern opportunities to renew the social contract between public institutions and inhabitants, take on a territorial approach to development, empower modern designs of utilization and production, and track a decrease in the utilization of natural assets. There are five lines of activity that can help oversee the improvement of the blockchain innovation in cities, whereas integrating it in particular areas that can be adjusted and connected to distinctive cities.

11.4.1 Citizenship and Democracy

One of the fundamental human rights is the right to identity, and a birth certificate is an official document for our existence [10]. The source of our rights and responsibilities is our identity, and it stands as the gateway to the services we require in the digital and physical world. The particular identification of inhabitants utilizing blockchain means they can be enrolled in an unmodifiable way, with anonymous information protection, whereas permitting a person-centered plan to be actualized. They can be involved in voting, which can help build trust in transparent governance to enable citizens to work together toward social cohesion through an experimental solid base. It ensures that voting is private and unalterable, as well as the transparency of the method. The innovation makes it less demanding to arrange elections; register and confirm voters; issue and check votes; and share, review, and approve the outcome.

11.4.2 Asset and Land Usage

Blockchain can help advantage one of the establishments of the city's economy: land and property development. It ensures that property proprietorship is recorded and records the transactions, commitments, and impositions of the real estate market with their corresponding funds. This permits an ecosystem to create around property value that incorporates all the partners: the common open, the individuals or associations that possess the arrive, related companies within the segment, certifying bodies, property appraisers, charge specialists, and arranging or urban organization specialists. Blockchain shows real estate property rights as computerized resources and keeps records of their enlistment and certification, any development or urban advancement rights, and the collateral utilized for financing. It also makes it simpler to access available information on the root, exchanges and commitments related to a property. Blockchain moreover makes it simpler to carry out real estate exchanges: holding, selling, buying, mortgaging genuine estate, and utilizing other financing instruments; verifying, certifying, evaluating, taxing, arranging and overseeing their value, keeping track of use, and transferring proprietorship.

11.4.3 Infrastructure and Services

There is a vivid potential for making markets for digital resource exchanges related to existing urban infrastructures, and this innovation is just beginning to be connected to city services. Blockchain will permit modern ecosystems of interactions among citizens, service providers, and city governments to be investigated. It offers the chance to build platforms that make our material quality of life better, provides business opportunities, and gives work for the city's occupants. In specific, modern energy markets are as of now in operation on the existing framework. The use of intelligent meters associated with the Internet, utilized as nodes for power utilization and generation in buildings, has made it conceivable to evaluate the cost, cogeneration, and payments comparing to person power use from the electric grid, in both bearings. As part of this process, blockchain makes it conceivable to automate the exchange of excess energy in any hub. It tracks each unit of electricity from generation to the point of utilization through the local electric network and matches each energy exchange with its equivalent financial exchange, rearranging the process and making it more productive [11].

11.4.4 Ecosystems of Values

The capacity to digitally register values and trace their roots, whether natural or cultural, makes it conceivable to direct new processes for creating and consuming goods. Blockchain is an "Internet of Value" and, if we expand its potential past its monetary esteem, we can encourage the exchange of nonfinancial values to start a productive dialogue that joins existential contrasts, and that reacts to the wants of communities with a run of particular interests, whatever they may be. If we need modern generation and utilization models for cities, we have first to determine which values we need to encourage and head toward. We, as of now, know the conventional financial operation for existing models. To improve on this, we ought to pay attention to trades of nonfinancial values in our communities. The blockchain exchange history can keep a record of any values that are appeared to be imperative to a community. If this includes the utilization of natural products, then the exchange history will show the roots of organic agriculture or the product's carbon footprint. If it comprises social products' utilization, the exchange history will show the trade of substance.

11.4.5 Government and Public Tenders

A permanent record of government activities will empower straightforwardness, whereas the capacity to check the record will lead to responsibility. Both angles will fortify modern models of governance and local independence, which can progress the way cities work through an arrangement of integrated urban instruments. With blockchain, restructuring authoritative frameworks will permit the most straightforward and most tedious capacities carried out by open specialists to be automated so that smart contracts can uphold citizens. Over time, and with experience being utilized in administration, smart contracts could begin to build programmed, decentralized hybrid frameworks (both human and advanced).

11.5 Conclusion

This chapter listed and explained some of the current challenges that are faced by the urban cities and suggested smart city as a possible candidate to help solve those challenges. Smart city solutions include smart energy, smart transportation, smart health, smart lighting, and smart waste management, among others to make the lives of the residents easier. With these smart systems in place, new technologies such as the blockchain innovation are needed to run them to ensure the security and privacy of the smart users. In addition, the transactions on these smart systems must be transparent, immutable, and decentralized. With that been said, the blockchain technology is seen as a suitable candidate that can provide previous functions. This chapter explained the function of blockchain technology in several smart city sectors as well as how the technology can be employed in smart city administration.

References

[1] S. Arora, "National eID card schemes: A European overview," Inf. Secure Tech. Rep., vol. 13, pp. 46–53, 2008.

[2] Getty Images, "Smart City Concept and Internet of Things Stock Vector Art & More Images of City 510687292 iStock," www.istockphoto.com, 2018. [Online]. Available: https://www.istockphoto.com/vector/smart-city-concept-and-internet-of-things-gm510687292-86357031. [Accessed: 16-Apr-2018].

[3] Adtell Integration, "Smart Cities Infrastructure – Adtell Integration – Total Communication Solution," adtellintegration.com, 2018. [Online]. Available: http://adtellintegration.com/smart-cities-infrastructure/. [Accessed: 16-Apr-2018].

[4] V. Albino, U. Berardi, and R. M. Dangelico, "Smart cities: Definitions, dimensions, performance, and initiatives," J. Urban Technol., vol. 22, no. 1, pp. 1–19, 2015.

[5] L. G. Anthopoulos and C. G. Reddick, "Smart City and Smart Government," in Proceedings of the 25th International Conference Companion on World Wide Web – WWW '16 Companion, 2016, pp. 351–355.

[6] J. Gao, K. O. Asamoah, E. B. Sifah, A. Smahi, Q. Xia, H. Xia, ... and G. Dong, "GridMonitoring: Secured sovereign blockchain based monitoring on smart grid," IEEE Access, vol. 6, pp. 9917–9925, 2018.

[7] S. Borlase et al., "Smart cities," in Smart Grids: Advanced Technologies and Solutions, Second Edition, 2017.

[8] M. Al-Hader and A. Rodzi, "The smart city infrastructure development & monitoring," Theor. Empir. Res. Urban Manage, vol. 4, no. 2[11], pp. 87–94, 2009.

[9] M. Angelidou, "Smart cities: A conjuncture of four forces," Cities, vol. 47, pp. 95–106, 2015.

[10] S. Zygiaris, "Smart city reference model: Assisting planners to conceptualize the building of smart city innovation ecosystems," J. Knowl. Econ., vol. 4, no. 2, pp. 217–231, 2013.

[11] P. Lombardi, S. Giordano, H. Farouh, and W. Yousef, "Modelling the smart city performance," Innovation, 2012.

Index

Note: Locators in *italics* represent figures and **bold** indicate tables in the text.